Studies in Systems, Decision and Control

Volume 311

Series Editor

Janusz Kacprzyk, Systems Research Institute, Polish Academy of Sciences, Warsaw, Poland

The series "Studies in Systems, Decision and Control" (SSDC) covers both new developments and advances, as well as the state of the art, in the various areas of broadly perceived systems, decision making and control–quickly, up to date and with a high quality. The intent is to cover the theory, applications, and perspectives on the state of the art and future developments relevant to systems, decision making, control, complex processes and related areas, as embedded in the fields of engineering, computer science, physics, economics, social and life sciences, as well as the paradigms and methodologies behind them. The series contains monographs, textbooks, lecture notes and edited volumes in systems, decision making and control spanning the areas of Cyber-Physical Systems, Autonomous Systems, Sensor Networks, Control Systems, Energy Systems, Automotive Systems, Biological Systems, Vehicular Networking and Connected Vehicles, Aerospace Systems, Automation, Manufacturing, Smart Grids, Nonlinear Systems, Power Systems, Robotics, Social Systems, Economic Systems and other. Of particular value to both the contributors and the readership are the short publication timeframe and the world-wide distribution and exposure which enable both a wide and rapid dissemination of research output.

** Indexing: The books of this series are submitted to ISI, SCOPUS, DBLP, Ulrichs, MathSciNet, Current Mathematical Publications, Mathematical Reviews, Zentralblatt Math: MetaPress and Springerlink.

More information about this series at http://www.springer.com/series/13304

Aboul Ella Hassanien · Aditya Khamparia ·
Deepak Gupta · K. Shankar · Adam Slowik
Editors

Cognitive Internet of Medical Things for Smart Healthcare

Services and Applications

Springer

Editors
Aboul Ella Hassanien
Department of Computer and Information Technology
Cairo University
Giza, Egypt

Aditya Khamparia
School of Computer Science and Engineering
Lovely Professional University
Phagwara, Punjab, India

Deepak Gupta
Department of Computer Science Engineering
Maharaja Agrasen Institute of Technology
Rohini, Delhi, India

K. Shankar
Department of Computer Applications
Alagappa University
Karaikudi, Tamil Nadu, India

Adam Slowik
Department of Electronics and Computer Science
Koszalin University of Technology
Koszalin, Poland

ISSN 2198-4182 ISSN 2198-4190 (electronic)
Studies in Systems, Decision and Control
ISBN 978-3-030-55832-1 ISBN 978-3-030-55833-8 (eBook)
https://doi.org/10.1007/978-3-030-55833-8

This Springer imprint is published by the registered company Springer Nature Switzerland AG
The registered company address is: Gewerbestrasse 11, 6330 Cham, Switzerland

Preface

This book begins with the basics of Internet of Medical Things (IoMT) and introduces the methodologies, processes, results and challenges associated with the same. Internet of Medical Things (IoMT) means the collection of medical devices and applications that connect to healthcare information systems via the Internet. Appearance of IoMT technology provides effective and reliable results to support healthcare services. It reduces the cost to consumers by improving clinical and quality services for patients. Due to evolution from 2018 to 2025, developments in IoT healthcare applications are to be sure ready to quicken as the Internet of Things is the key part in the advance change of the healthcare business and different partners are increasing their determinations. The main reason behind the success rate of IoMT is having the capability to help, monitor, inform not only caregivers with usage of wearable devices or telematics but provides healthcare providers actual data to identify issues before they become critical. This book will focus on involvement of IoMT-driven intelligent computing methods, state of the arts, novel findings and recent advances in medicine and health care due to new technologies and faster communication between users and devices. This is an exciting and emerging interdisciplinary healthcare-related area in which a wide range of theory and methodologies are being investigated and developed to tackle complex and challenging problems. Application-oriented papers are expected to contain cloud IoMT analysis, machine learning, computer vision and deep learning-enabled evaluation of the proposed solutions. This book further illustrates the possible challenges in its applications and suggests ways to overcome them. The topic is wide in nature, and hence, every technique and/or solution cannot be discussed in detail. The primary emphasis of this book is to introduce different IoMT-assisted remote healthcare monitoring applications, challenges and concepts to researchers, students and academicians at large.

Objective of the Book

The main aim of this book is to provide a detailed understanding of IoMT-assisted applications with involvement of distinct intelligent computing methods and optimized algorithms in the field of computer science. This book will endeavor to endow with significant frameworks, theory, design methods and the latest empirical research findings in the area of IoMT to foster healthcare sector that can be put to good use.

Organization of the Book

This book is organized into 13 chapters with the following brief description:

1. **A Review of Applications, Security and Challenges of Internet of Medical Things**

The Internet of Medical Things (IoMT) plays a crucial role in enhancing the quality, efficiency and effectiveness of its products in the healthcare field. The chapter discusses cross-review of all IoMT chosen papers with some latest research material and articles combined. This could help researchers consider previous applications, problems, challenges and threats in the healthcare field. The presented work also includes an overview of the IoMT design and how cloud storage technology supports healthcare applications.

2. **IoT Enabled Technology in Secured Healthcare: Applications, Challenges and Future Directions**

This chapter introduces Internet of Health Things with wearable healthcare systems in detail and shows the interassociation of interaction allowed medical gadgets and their combination to broader scale networks of health care to enhance the health of patients, because of this sensitive behavior of systems related to health.

3. **A Comparative Analysis of Image Denoising Problem: Noise Models, Denoising Filters and Applications**

Noise generates maximum critical disturbances as well as touches the medical image quality and ultrasound images in the field of biomedical imaging. This chapter highlights the comparative analysis of image denoising problems with the help of Gaussian, Weiner and mean filter using PSNR and related metrics.

4. **Applications and Challenges of Cloud Integrated IoMT**

This chapter focuses on different aspects including the strengths, weakness, prospects and challenges of the IoMT integrated cloud computing. It also describes a framework for IoT healthcare network (IoThNet) is presented which illustrates how hospitals at access layer can collect user information at data persistent layer.

A description is also provided on how IoMT can help support different diseases with the help of sensors, for example, glucose, pulse, temperature, blood pressure, heart rate, force, etc.

5. **Optimal SVM Based Brain Tumor MRI Image Classification in Cloud Internet of Medical Things**

This chapter discusses the new detection and diagnosis model for brain tumor. The proposed gravitational search algorithm with genetic algorithm (IGSAGA) model is applied for filtering the features, and optimal support vector machine (SVM) model is applied for classification processes. The results are validated using a benchmark BRATS dataset, and the experimental outcome indicated the supremacy of the projected model.

6. **An Effective Fuzzy Logic Based Clustering Scheme for Edge-Computing Based Internet of Medical Things Systems**

This chapter presents an effective fuzzy logic-based clustering technique for IoMT applications. The presented FC-IoMT technique selects the cluster heads (CHs) based on five input parameters, namely energy, distance, delay, capacity and queue. The proposed model has undergone extensive validation, and the results ensured the superior results under several measures.

7. **Automated Internet of Medical Things (IoMT) Based Healthcare Monitoring System**

This chapter introduces an IoMT-based automated healthcare system for remote patient that helps physicians and their connections. To predict the disease, the system conducts mechanical training using the CHAID algorithm (chi-square automatic interaction detection performs multi-level splits when computing classification trees) and generated multiple distributions during tree sorting to sort this data. This structure allows the doctor to intervene immediately to help patients with unusual health problems.

8. **Deep Belief Network Based Healthcare Monitoring System in IoMT**

This chapter describes how the Internet of Things can be used for machine learning in health care. Based on the deep belief network (DBN) and IoMT, i-NXGeVita can detect normal and abnormal heart rates and classify various defects. The system achieves the accuracy of 97% for the proposed healthcare monitoring system.

9. **An IoMT Assisted Heart Disease Diagnostic System Using Machine Learning Techniques**

In this chapter, various machine learning algorithms have been implemented to predict the heart diseases using IoMT. 88.59% accuracy was obtained by logistic regression using majority voting which is far better than the existing system techniques.

10. **QoS Optimization in Internet of Medical Things for Sustainable Management**

This chapter demonstrated QoS optimization in IoMT for sustainable management in wireless sensor networks. The result emphasized that hybrid wireless mesh networks (HWMNs) performed more efficiently when compared with ad hoc on-demand distance vector (AODV) and secured AODV (SAODV) routing protocols.

11. **An Intelligent Internet of Medical Things with Deep Learning Based Automated Breast Cancer Detection and Classification Model**

The chapter presented an intelligent IoMT-based breast cancer detection and diagnosis using deep learning model. The proposed model performs a set of processes, namely preprocessing, K-means clustering-based segmentation, local binary pattern (LBP)-based feature extraction and deep neural network (DNN)-based classification. The experimental results ensured the superior performance of the LBP-DNN model with the maximum sensitivity of 71.64%, specificity of 75.87% and accuracy of 70.53%.

12. **Internet of Medical Things (IoMT) Enabled Skin Lesion Detection and Classification Using Optimal Segmentation and Restricted Boltzmann Machines**

The proposed chapter introduces a new IoMT-based skin lesion detection and classification model using optimal segmentation and restricted Boltzmann machines (RBMs), named OS-RBM model. The proposed OS-RBM model involves a series of steps, namely image acquisition, Gaussian filtering (GF)-based preprocessing, segmentation, feature extraction and classification. The experimental outcome ensured the effective classification performance of the OS-RBM model with the maximum sensitivity of 96.43%, specificity of 97.95% and accuracy of 95.68%.

13. **An IOT Based Medical Tracking System (IMTS) and Prediction with Probability of Infection**

The chapter discussed an IoT-based medical appliance called IMTS which works on various integrated handheld medical devices such as pulse sensor, thermal sensor and oximeter followed by a data analytic experiment over the data received from all such devices corresponding to any patient. The data and result show that the proposed system is a handy tool at domestic level and carrying a good concept.

Giza, Egypt — Aboul Ella Hassanien
Phagwara, India — Aditya Khamparia
Rohini, India — Deepak Gupta
Karaikudi, India — K. Shankar
Koszalin, Poland — Adam Slowik
June 2020

Contents

About the Editors

Dr. Aboul Ella Hassanien is Founder and Head of the Egyptian Scientific Research Group (SRGE). He has more than 1000 scientific research papers published in prestigious international journals and over 50 books covering diverse topics such as data mining, medical images, intelligent systems, social networks and smart environment. He won several awards including the Best Researcher of the Youth Award of Astronomy and Geophysics of the National Research Institute, Academy of Scientific Research (Egypt, 1990). He was also granted a Scientific Excellence Award in Humanities from the University of Kuwait for the 2004 Award and received the superiority of scientific—University Award (Cairo University, 2013). Also, he honored in Egypt as the best researcher in Cairo University in 2013. He also received the Islamic Educational, Scientific and Cultural Organization (ISESCO) prize on Technology (2014) and received the State Award for Excellence in Engineering Sciences 2015. He was awarded the Medal of Sciences and Arts of the first class by the President of the Arab Republic of Egypt, 2017. He was awarded the International Scopus Award for the meritorious research contribution in the field of computer science (2019).

Dr. Aditya Khamparia has expertise in teaching, entrepreneurship and research & development of 8 years. He received his Ph.D. degree from Lovely Professional University, Punjab, in May 2018. He has completed his M.Tech. from VIT University and B.Tech. from RGPV, Bhopal. He has completed his PDF from UNIFOR, Brazil. He has around 65 research papers along with chapters including more than 15 papers in SCI indexed journals with cumulative impact factor of above 50 to his credit. Additionally, he has authored and edited five books. Furthermore, he has served the research field as Keynote Speaker/Session Chair/Reviewer/TPC Member/Guest Editor and many more positions in various conferences and journals. His research interests include machine learning, deep learning, educational technologies, computer vision.

Dr. Deepak Gupta is an eminent academician and plays versatile roles and responsibilities juggling between lectures, research, publications, consultancy, community service, Ph.D. and postdoctorate supervision, etc. With 13 years of rich expertise in teaching and two years in industry, he focuses on rational and practical learning. He has contributed massive literature in the fields of human-computer interaction, intelligent data analysis, nature-inspired computing, machine learning and soft computing. He has served as Editor in Chief, Guest Editor and Associate Editor in SCI and various other reputed journals. He has completed his postdoc from Inatel, Brazil, and Ph.D. from Dr. A.P.J. Abdul Kalam Technical University. He has authored/edited 37 books with national-/international-level publisher (Elsevier, Springer, Wiley, Katson). He has published 132 scientific research publications in reputed international journals and conferences including 62 SCI indexed journals of IEEE, Elsevier, Springer, Wiley and many more. He is the convener and organizer of 'ICICC' Springer conference series.

Dr. K. Shankar is currently Postdoc Fellow in the Alagappa University, Karaikudi, India. Collectively, he authored/co-authored over 50+ ISI journal articles (with total impact factor 150+) and 100+ Scopus indexed articles. He has guest-edited several special issues at many journals published by SAGE, TechScience, Inderscience and MDPI. He has served as Guest Editor and Associate Editor in SCI, Scopus indexed journals like Elsevier, Springer, IGI, Wiley & MDPI. He has served as Chair (program, publications, Technical Committee and track) on several international conferences. He has delivered several invited and keynote talks, and reviewed the technology leading articles for journals like Scientific Reports—Nature, the IEEE Journal of Biomedical and Health Informatics, IEEE Transactions on Reliability, the IEEE Access, the IEEE Internet of Things, Big Data Research, Human-centric Computing and Information Sciences, and the IEEE Internet of Things. He authored/edited conference proceedings, chapters and two books published by Springer. He has been a part of various seminars, paper presentations, research paper reviews, and convener and a session chair of the several conferences. He displayed vast success in continuously acquiring new knowledge and applying innovative pedagogies, has always aimed to be an effective educator and has a global outlook. His current research interests include healthcare applications, secret image sharing scheme, digital image security, cryptography, Internet of Things and optimization algorithms.

Adam Slowik (IEEE Member 2007; IEEE Senior Member 2012) received the B.Sc. and M.Sc. degrees in computer engineering and electronics in 2001 and the Ph.D. degree with distinction in 2007 from the Department of Electronics and Computer Science, Koszalin University of Technology, Koszalin, Poland. He received the Dr. Habil. degree in computer science (intelligent systems) in 2013 from the Department of Mechanical Engineering and Computer Science, Czestochowa University of Technology, Czestochowa, Poland. Since October 2013, he has been Associate Professor in the Department of Electronics and Computer Science, Koszalin University of Technology. His research interests include soft computing, computational intelligence and, particularly, bio-inspired optimization algorithms and their engineering applications. He is a reviewer for many international scientific journals. He is Author or Coauthor of over 80 refereed articles in international journals, two books and conference proceedings, including one invited talk. He is Associate Editor of the IEEE Transactions on Industrial Informatics. He is Member of the program committees of several important international conferences in the areas of artificial intelligence and evolutionary computation. He was a recipient of one Best Paper Award (IEEE Conference on Human System Interaction—HSI 2008).

A Review of Applications, Security and Challenges of Internet of Medical Things

Shashank Kumar, Arjit Kaur Arora, Parth Gupta, and Baljit Singh Saini

Abstract The Internet of Medical Things (IoMT) relates to the interconnectedness between connectivity-enabled medical equipment and their incorporation into larger health networks in order to enhance the health of patients. The Internet of Medical Things plays a crucial role in enhancing the quality, efficiency and effectiveness of its products in the healthcare field. While Internet of Things takes together many fields, but our emphasis is on IOT's work impact in the area of healthcare. This paper consists of a cross-review of all those carefully chosen papers with some latest research material and articles combined. This review should help researchers consider previous applications, problems, challenges and threats in the healthcare field. This paper also includes an overview of the IoMT design and how cloud storage technology supports healthcare applications. We assume that this review can be helpful to researchers and professionals in the area, allowing them to appreciate the immense possibility of IoT in the medical world.

1 Introduction

Throughout developed countries the health system is evolving rapidly, with a significant increase in the life expectancy in the twentieths century [1]. The health systems of these countries are also getting pressured by the increase in the chronic illnesses

S. Kumar · A. K. Arora · P. Gupta · B. S. Saini (✉)
CSE, Lovely Professional University, Punjab, India
e-mail: baljitsaini28@gmail.com

S. Kumar
e-mail: theshashank6870@gmail.com

A. K. Arora
e-mail: arjitkaurarora8@gmail.com

P. Gupta
e-mail: guptaparth118@gmail.com

A. E. Hassanien et al. (eds.), *Cognitive Internet of Medical Things for Smart Healthcare*, Studies in Systems, Decision and Control 311,
https://doi.org/10.1007/978-3-030-55833-8_1

of the citizens [2]. Indeed, life expectancy in developed countries in the twentieth century has increased by about 30 years. The number of older adults is therefore increasingly increasing [1]. In fact, the proliferation of chronic disorders has placed a lot of strain on healthcare services across the world because of a shortage of resources [2]. In fact, IoMT incorporates both the functionality and protection of conventional medical equipment with modern Internet of Things (IoT) advanced, popular, and portable technology. Besides being perfectly robust to deal with a wide range of diseases requiring myriad inspection requirements, it also has the ability to tackle the problem of diseases that persists for a longer period of time by being able to handle several systems configured for a variety of individuals. Indeed, IoMT also offers a solution to other issues such as systematic real-time monitoring of patients in their everyday lives, as opposed to telemedicine online programs that focus solely on home care [2]. IoMT is not only a connection between a vast number of personal medical instruments but often amongst medical equipment and the ones who supply health experts, or commercial firms. IoMT's growth is largely attributed to the expanded usage and creation of wired and portable medical tools, taking with them both innovative technologies and numerous challenges [3]. The Internet of Medical Things (IoMT), sometimes described as Healthcare IoT, refers to medical device collection as well as software linked via networks. Many healthcare professionals use IoMT software to optimize diagnosis, control illnesses, reduce delays, enhance customer service, administer medications and eliminate prices. Market research indicates that the IoT healthcare market would be expected to hit $117 billion until 2020 [1, 4]. In this paper we will review some of the most important points in the field of IoMT. Papers reviewed are themselves review papers in different subsets of IoMT. This paper consists of a crux review of all those carefully selected papers with some new content by some recent papers as well.

2 Applications of IoT in Healthcare

The main factors of deaths of women between the ages of 20 and 59 is breast cancer, impacting 2.1 million females per year According to WHO, in the year 2018, it was recorded that 627,000 people died from breast cancer—about 15% of all women's cancer deaths discussed in [5]. This can be reduced by getting effective care and proper disease treatment. Various artificial intelligence models act as an useful feature in the application of technology to that of the Internet of Medical Issues (IoMT) which can be used to further detect and treat malignant cells in breast cancer.

IoT has made life more convenient than ever. With advancement in the IoT sector, care for patients is increasing, healthcare is becoming cheap, outcomes of patients have improved, any type of disease can be detected in real time, one's life quality has improved and user end experience is improving. One's main aim to sustain in this world is health and long lives, that too is achieved by this sector. It is also used to efficiently monitor most of the diseases and its prevention. Any sudden drop in the health of patients will show an automatic alert to different healthcare officials,

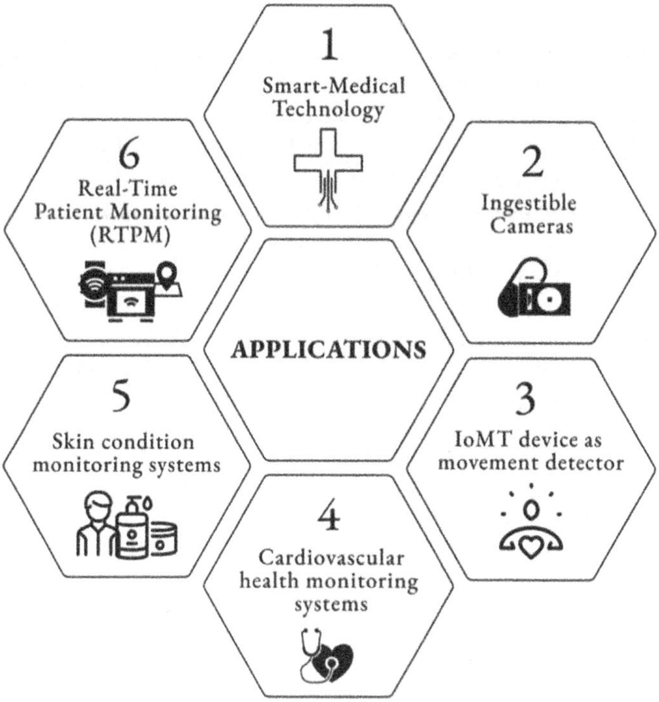

Fig. 1 Applications of IoMT

saving resources of other IoT based devices [6, 7]. Every patient's medication is immediately imitated to the family members [8]. In the IoT sector, one of the most crucial factors that help IoT to grow is User Experience (UX). It provides simplicity, affordability and easy to use devices with the least amount of instructions needed to explain the working of the device [9].

With this, it also becomes easy for doctors to keep a record of many patients in a single place [10]. Time and energy consumption of doctors reduces [11]. Strong and healthy diets recommended by dieticians or any medical diet based Machine learning/Artificial Intelligence model helps the patient to prevent themselves from further disease, and also it improves the well being. But, medical professionals fail to thoroughly grasp the reasoning of the patient-dietitian for the recommended method [12]. The application areas can be broadly categorized as shown in Fig. 1.

2.1 Smart-Medical Technology

This includes smart medical equipment and kits that are currently being deployed. These are currently used by paramedics to provide immediate help to patients who are in urgent need of medical care and assistance. One example is the use of medical drones to perform such a task [13]. Medical drones were originally introduced to respond to emergencies related to patients suffering from cardiac arrests [14], since these drones are the fastest to arrive at the emergency scene. The drones would be directed to fly to specific destinations, which saves time and as such, saves lives since paramedics might end up stuck in traffic, and may not be able to respond as quickly as needed. This encourages the reliance on smart medical robots [15] to perform surgical operations within a hospital setting. Virtual/Augmented Reality and Artificial-Intelligence (AI)-based medical technologies were also employed for various medical purposes. AI-based medical technologies are also being used that ensure a higher accuracy rate [16]. These models include biochemical interactions [17], such as IBM Watson and Gene Network Sciences (GNS) Healthcare AI systems [18] used to search for the right cancer treatment [19].

2.2 *Ingestible Cameras*

These are cutting-edge and cost-effective capsules that can be swallowed (in-vivo/in-vitro) by a patient to provide internal-organ real-time visual monitoring for early detection of chronic diseases and cancer [20]. Many ingestible devices were presented including Swallow-able data recorder capsule medical device [21], ingestible endoscopic optical scanning device for endoscopy [22], and ingestible hydrogel device [23]. Ingestible devices rely on an X-ray or camera capsule, a tracking/recording system and the diagnostics toolkit for evaluation.

2.3 *Real-Time Patient Monitoring (RTPM)*

This is a new evolving trend among the new generation, including millennials, due to their heavy reliance on smart devices as a key part of their daily lives [24]. In fact, RTPM is used to ensure a real-time, cost-effective remote consistent monitoring depending on the sensors linked to the patient's body, either through a homecare telehealth system [25, 26] or telecare monitoring systems [27, 28]. This may include monitoring fitness level, glucose level, respiration rate, and heart rate, etc. Many new RTPM trends are now available including, Apple Watch app that monitors depression, Apple's Research Kit and Parkinson's disease and ADAMM intelligence Asthma Monitoring [29, 30].

2.4 Cardiovascular Health Monitoring Systems

These systems use heart rate and pulse wave velocity of the patient. It can help in early peripheral arterial disease (PAD) recognition by measuring Pulse Wave Velocity (PWV) with the help of Body Cardio Scale. For example—(Nokia, Helsinki, Finland) [31].

2.5 Skin Condition Monitoring Systems

These systems determine the wounded circumference and healing progress. Authors in [32] proposed to make a forum for mobile telemedicine with the help of smartphone. They tested the platform's viability as well as usability based on simulating testing by 10 experts, who used the experimental mobile app to remotely inspect a diabetic foot. This platform can remotely classify wounds as well as the amputation risks with an accuracy of 89% on average.

2.6 IoMT Device as a Movement Detector

These devices are for the immobile patients. It has become important for these groups of people to trace their movements. And that is why smart watches and sensors are mounted on the patient's clothes, bed, or body to track their activities. It will also assist in monitoring involuntary gestures and provide deeper insight into effective medical control. Data collected from these devices is protected from end to end, granting patients more protection.

3 IoT Healthcare Security and Privacy

Efficient approaches have been demonstrated in home telemedicine system [33] to minimize overloading of healthcare services and to decrease the cost of healthcare. As so many IoT components wirelessly transmit and receive data which puts IoMT at risk of security breaches of the wireless sensor network (WSN) [34]. IoMT solutions also provide software for running, tracking, and managing these. Application risks that include authentication and authorization breaches, along with the general protection and functionality of the program, there is also concern [35]. Because of the vulnerability of all the medical systems, it has become important to handle in a heterogeneous fashion. Many of the MTs which are linked with the internet that ensures reliability. Hence, we require a versatile layered architecture. The five-layer

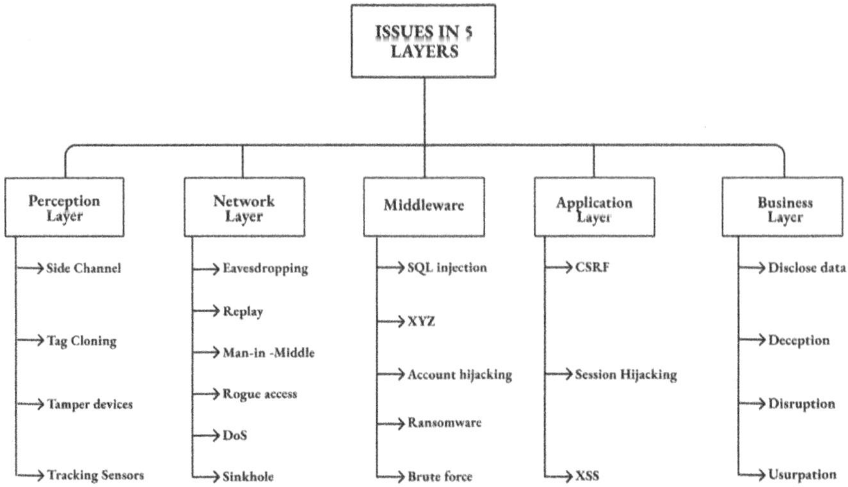

Fig. 2 Layer-wise security issues

IoT model mentioned in [36, 37] presents a layered architecture that provides unique functionality for each layer. Layers are as follows:

1. Perception Layer: Here all the data that includes body temperature, heart rate, etc. which utilizes the physical devices like sensors and transmits the data to the network i.e. network layer.
2. Network Layer: This layer searches and delivers the content and routes the content from the source to the destination with the use of network addressing.
3. Middleware: This layer manages the processing and retrieval of data that is retrieved from the perception layer instruments i.e. sensors, the detection of resources and the management of access to apps.
4. Application Layer: With the help of this layer all the users link to IoMT devices through a middleware layer.
5. Business Layer: It controls the market rationale of healthcare providers and manages the life-cycle of the company operation, including tracking, controlling and improving business processes.

These layers and the issues are summarized in Fig. 2

3.1 Perception Layer Issues

1. Side Channel: Attackers can use a variety of side-channel methods, for example, Analysis of pacing and resource usage and help to track the electromagnetic (EM) activity across medical instruments for the extraction of data [38].

2. Tag Cloning: Here the attacker uses the data obtained through an effective side channel attack or it duplicates the data through the existing tag [39].
3. Tampering the devices: As the sensors can be amenable, hence, any attacker can physically interfere with sensors and incompletely or fully interrupt or control the performance. Attackers can connect to these IoMT devices in order to disable the medical system containing a USB port [40]. Also, the attackers may also interfere with computers by making use of certain software bugs to install malicious software, which gives the attacker the ownership over the particular device [41].
4. Tracking Sensors: Utilizing insecure tools, attackers can be comfortable with patient positions or GPS data spoof [42]. Vulnerabilities in cellular network signaling system No. 7 (SS7) can compromise MT [43].

3.2 Network Layer Issues

1. Eavesdropping: Attackers identify and decrypt the necessary equipment with which they can effectively capture details sent by all the hardware devices.
2. Replay: In this, the attacker replays the authentication message that has previously been shared among legal users. For example, OneTouch Ping insulin pump is considered susceptible as it is in need of safe contact procedure [44].
3. Man-in-the-Middle: An unauthorized party can get into data using a backdoor and replay and change the communications of legal parties [45].
4. Rogue access: Here, a fake portal is established within the wireless network which scopes to enable legal access to the system and, in turn, to intercept data [46].
5. Denial of Service (DoS): Here, the attack disrupts the availability of medical equipment and their services by flooding the network with service requirements. Distributed DoS (DDoS) is a more sensitive variant of DoS that needs several nodes well into the floods, rendering this harder and harder to find. In 2016, large-scale DDoS attacks using IoT instruments were successful due to the wide-scale deployment of IoT devices.
6. Sinkhole: In such type of attack, a malicious node draws traffic by providing higher connection efficiency (e.g. ads for fake routes).

3.3 Middleware Layer Issues

1. Cross-site request forgery (CSRF): Here, the attack is popular in RESTful IoT networks. The CSRF tricks the end user into taking action on a compromised program without the consent of the consumer. According to the HP survey of many IoT ventures, 60% of all cloud-connected IoT systems have cross-site request forgery (CSRF) and client hijacking vulnerabilities [47].
2. Session Hijacking: It is also growing in RESTful IoT based systems. In this intruder, the session and data access may be taken over

3. Cross-site scripting (XSS): XSS also exploits RESTful IoT programs by adding side scripts to circumvent site page access restrictions.

3.4 Applicatsion Layer Issues

1. SQL injection: The SQL injection intrusion occurs anytime an attacker intends to access a client-linked backend database by inserting a malformed SQL code.
2. Efficient SQL injection attacks may negotiate confidential health details or change vital information [48].
3. Account hijacking: The intruder can steal the identity by intercepting a packet while the user authenticates it.
4. Ransomware: Ransomware encrypts confidential information and requests a large fee for recovery. The danger begins with a single device, and afterwards spreads to the whole network. Attackers can encrypt personal details, such as patient records, and hold the decryption key in exchange for money.
5. Brute force: This is simply attempting to infer inputs including passwords by checking all potential combos [37].

3.5 Business Layer Issues

1. Disclosure of information: Unauthorized access to private details such as health reports breaches the security of the IoMT program. Approximately 38% of accidents to healthcare were of this kind in 2015 [49].
2. Deception: Attacks such as man-in-the-centre and sinkhole will contribute to knowledge manipulation. About 58% of health organizations don't have a system in place to correct such erroneous information [50].
3. Disruption: Disruption of correct protocol or exposure to vital information decreases the consistency of the system and may lead to life-threatening effects [51].
4. Usurpation: Unauthorized exposure to other parts of the network by attacks, like replay, malware injection and sinkhole, impacts the health of medical equipment [52].

SYN Flooding Attacks or "half-open" attacks mostly threaten high-capacity IoMT systems as they rely on Transmission Control Protocol (TCP) infrastructure to interact (i.e. email/web servers) [53]. The purpose of this attack is to trigger a medical system to crash by draining the memory buffer of the e-Healthcare system and render unsafe links accessible for further attacks. Black Nurse Attacks are highly successful low bandwidth (15–18 Mbit/s) ICMP attacks targeting high load Central Processing Unit (CPU) firewalls via denial of service attacks [52]. Birthday Attacks are often attributed to users depending on faulty hashing algorithms, where two separate passwords that have the same hash. This vulnerability can be abused easily to obtain unauthorized access to any medical system. A suggested combination of hash functions was presented at [54]. However, frameworks with Stable Hash Algorithms (e.g. SHA-3 and SHA-512) remain the best protection against these attacks.

There are several other security issues that are discussed in [55] which are given as follows:

5. **Authentication**: IoT authentication is very difficult, since heterogeneous network authentication is involved. Items (sensors) have to be detected and authenticated prior to entering the network. It needs a global unique identification token for each individual in the network.
6. **Confidentiality**: To be certain that unauthorized users cannot access health data. Confidential communications should be kept private from the eavesdroppers.
7. **Self-Sustaining**: When a remote patient (RPMN) medical system fails then the other healthcare systems must be able to provide a minimum level of protection.
8. **Fault Tolerance**: The system should come up with applicable security providers with the functionality in the event of computer failure or breach.
9. **Flexibility**: The device should be flexible enough to defend the network even if any IoT nodes are compromised.
10. **Data Freshness**: Recent (fresh) data must be available to Remote Patient Monsitoring networks to operate efficiently with nodes. Of example, the most recent ECG readings are required to evaluate the heart functioning of every patient Consultant.
11. **Anonymity**: A lot of patients want their identity to remain anonymous inside RPMS.
12. **Liability**: Remote Patient Monitoring systems should establish liability for any abuse, failure, theft or unusual event.
13. **Trust**: Users or patients need guarantees to prevent any personal and medical information theft in Remote Patient Monitoring systems.

4 Security Measures

IoMT suffers from different issues and challenges such as the lack of security and privacy measures, in addition to the necessary training and awareness. We're going to review the existing security solutions which are discussed in [56]. The solutions are classified as cryptographic and non-cryptographic. As there is a rise of digital healthcare v4.0 era, exhibiting a trade-off between the security level and the system performance for IoMT act as a major security measure. We're providing a security solution, which is divided into five different layers to detect and prevent attacks, in addition to reducing the damage of these known attacks and preserving the patients' privacy. The devices are capable of constantly tracking the patient's health in real-time, offering them physical flexibility and mobility. This includes monitoring fitness level, glucose level, respiration rate, and heart rate, etc. Hence, there is no need to keep them in hospitals. The main issue is that many IoMT devices are prone and vulnerable to cyber-attacks simply because medical devices are either poorly secured

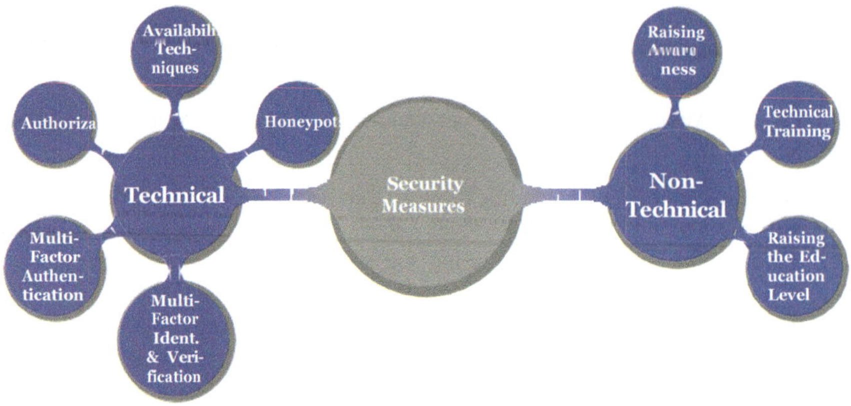

Fig. 3 Security Measures

against potential adversaries or not secure at all. This is the main challenge that IoMT faces i.e. preserving the patient's privacy without degrading the security level. No development in this field would hinder the wider deployment of IoMT. Overcoming the rising security issues and challenges is a challenging task. However, mitigating them can be achieved by implementing multiple security measures, some being technical and some non-technical (refer Fig. 3).

1. Non-Technical Security Measures: Theses measures include training of the staff and also securing the medical information of the patients. These measures can be applied according to the need. Training the medical and IT staff could be accomplished in three different ways:

   a. Raising Awareness: It is mandatory to aware the staff, especially the IT staff as they can know and identify an occurring attack from a normal network. It helps them to find the possibility to assess the likelihood and impact of a risk. Once risk is identified, it is also essential to explain how to mitigate it and use the right security measures to deal with any threat and reduce its risk. For this, we need to spread technical awareness.
   b. Technical Training: Besides raising awareness, people should know how to mitigate the risks and handle the attacks. It is equally important to start training the medical staff and employees of the IT department, right after the teaching phase. This includes 7 phases:

      - Identification Phase where the IT is capable of identifying a suspicious behaviour from an abnormal behaviour.
      - Confirmation Phase that is based on the ability to confirm that an attack is occurring.
      - Classification Phase that is based on the ability to identify the type of the occurring attack.

- Reaction or Responsive Phase is based on the ability of the Computer Emergency Response Team (CERT) to quickly react to a given attack using the right security defensive measures and prevent an attack from escalating.
- Containment Phase is based on containing the attack incident and overcoming it.
- Investigation Phase is the implementation of forensic evidence where an investigation process takes place to identify the cause of the attack, its impact and damage.
- Enhancement Phase is based on learning from the lessons of previous attacks.

c. Raising the Education Level: Education is a must in the IT department. This is based on teaching and educating cyber-security and IT staff the necessary techniques to classify each attack and what it targets like confidentiality, integrity and authentication.

2. Technical Security Measures: We discuss here the technical security measures that should be put in place to ensure an end-to-end secure IoMT system.

   a. Multi-Factor Identification and Verification: To prevent any possible unauthorized access to IoMT systems, it is necessary to ensure a robust identification and verification mechanism. The best solution is to have a biometric system. There is also a need for a database system to store the identities. There are several biometric techniques to achieve this, which is further divided into physical and behavioural techniques.

      - Physical Biometric Techniques: A biometric that is based on a physical trait of an individual. This includes facial recognition, retina scan, or iris scan.

        – Facial Recognition: Facial identification is the method of recognizing or confirming a person's identity through his or her features. It records, analyzes and contrasts trends which are focused on the facial features of the individual. The facial recognition process is an important phase in identifying and finding human faces in photographs and recordings.
        – Iris Scan: Iris detection or iris scanning is the practice of making a high-contrast image of a person's iris utilizing visible and near-infrared illumination.Thanks to its capacity to obtain precise and correct measurements, proven to be important for both recognition and verification purposes.
        – Retina Scan: A check for retinal identification is focused on examination of the area of the blood vessels behind the human eye. This has proved to be a very accurate and reliable test.

      - Behavioral Biometric Technique: Behavioral biometric technique that can be used for both identification and verification phases is the hand geometry. Techniques of behavioral biometric authentication include keystroke analysis, gait identification, speech ID, and features of mouse use, signature identification and cognitive biometrics. Generally, the precision of the biometric modality is determined on the basis of reciprocal errors consisting of both the false acceptance rate (FAR) as well as the false rejection rate (FRR).

b. Multi-Factor Authentication Techniques: Authentication is classified as the first line of defense that authenticates the source and destination alike. In fact, authentication can be a single-factor authentication that only relies on a password as the only security measure, which is not preferable. It can also be a two-factor authentication that relies on another security measure aside from the password in order to access a given system. Finally, it can be a multi-factor authentication where a third security mechanism is required in order to access a system. Therefore, authentication plays a key role in providing security for the accessible resources on a given network.
c. Authorization Techniques: An assigned authorization must be based on offering the least privilege. Hence, the Role Based Access Control i.e. the RBAC model is adopted. The model offers the least privilege for a given medical staff or employee to perform a given task with the least permissions and functionalities to accomplish a specific task. **T-Role-Based Access**: It is mainly designed for cloud computing environments, especially where medical data is stored [57]. T-RBAC also stands for Temporal Role Based Access Control, and can be spatio-temporal [58], intelligent, and generalized. It is also capable of validating any needed access permission for any medical user according to the assigned role and tasks.
d. Availability Techniques: Maintaining of servers is a must for easy flow of data. Maintaining the server's availability requires the implementation of computational devices that act as backup devices, along with verified backup and Emergency Response Plans (ERP) in case of any sudden system failure. **Against Jamming**: Jamming is the intentional intrusion, obstruction or interfering with licensed wireless communications. In static networks much jamming identification and countermeasure is planned and tested. The issue of ant jamming is more difficult in a mobile network system where jammers can switch and trigger jammer detection and localization algorithms to malfunction.
e. Honeypots: Honeypot systems are really useful when it comes to detecting attackers, their targets, tools and used methods. A honeypot is a network-attached device set up as a decoy for preventing, deflecting, or researching intrusion attempts to obtain unauthorized access to information systems.

5 IoMT Design

Two design approaches are defined in order to cope effectively with the challenges of reliability, validity, stability, modeling, testing, robustness and validation of IoMT.

5.1 Model-Based Development

We addressed contributions dealing with all facets of medical cyber-physical systems (CPS) and, more precisely, physiological models, since robust systems models have increased functionality and health. With the help of networking digital models, physiological modeling can easily be extended to amalgamate various IoMT devices [3]. In order to design a model of hybrid systems, first one should have a basic knowledge on continuous simulation systems modeling. Continuous simulation can be defined as a computer model of a physical system that continuously tracks system response based on a set of equations typically involving differential equations. Then, the modeling can further be used in discrete-time systems using a fault-tolerance approach [59]. This is the foundation for many more developed versions of continuous time, discrete time or mixed structures.

5.2 Service Layer Based IoMT Design

There is another unexplored approach based on service layer, where the design of IoMT systems is considered from a service oriented perspective. The quality-of-service (QoS) specifications, methods and templates are the key focus of the literature utilizing a service-oriented method to the architecture of IoMT. Subtle network controllers are then used to assure the required IoMT QoS on a large scale, combining all patient and clinical facilities [60]. Generally, there is different perspective from which QoS requirements of (Cyber-Physical Systems) CPS are well-developed:

1. Net Qos Criteria for CPS [61]
2. The QoS management architecture of both CPS application as well as the CPS framework [62].
3. The UML (Unified Modeling Language) is a focused QoS modeling for CPS [63]. Still on the issue of QoS management, QoS-enhancing middleware by considering resource managers and network resources has been created [64].
4. A system has been developed to ensure sufficient QoS by network resource management [65]. Numerous model-based design approaches have been suggested for the efficient development of service-oriented cyber-physical systems: software research and design languages have been used for designing architectures [66], including the expansion of the OWL-S to enable the advancement of service-based CPS prototypes [67]. Another subject of interest to IoMT-based applications is the implementation of suitable middleware, which provides a degree of complexity to system management by providing a standardized framework for linked artifacts. The service-oriented middleware design has been described utilizing either minimal-level C structures [68] or high-level XML definition for entity incorporation into huge-scale IPv6 networks [69].

6 Role of Cloud in IoT Healthcare Systems

The use of cloud technology in healthcare systems to store medical records is discussed in [70], which also reviews that cloud can be counted as a complete sector. With the help of cloud technology, we came to know how a huge database could provide various insights for health data analysis and determine patterns in various types of diseases, which is considered in [71]. Every work considered above provides valuable understanding in the cloud technology field. But there are few articles that discuss all of the prospects of cloud in healthcare systems based on wide body-area networks and the Internet of Things. In this section, we can fill this gap by presenting recent works regarding healthcare systems based on cloud technology. The use of cloud in IoMT can be broadly categorized into three categories as shown in Fig. 4.

6.1 Cloud for Healthcare

In recent years, after various researches has provided numerous insight of the benefits of cloud storage technology in healthcare applications, we came to know that the benefits that can be provided by cloud storage technologies are derived from mainly three services in healthcare sector:

14. Software as a Service (SaaS)—provides online software to healthcare providers that will enable them to work with health data or perform other relevant tasks.
15. Platform as a Service (PaaS)—provides tools for virtualization, networking, database management, and more.
16. Infrastructure as a Service (IaaS)—provides the physical (hardware) infrastructure for storage, servers, and more [72].

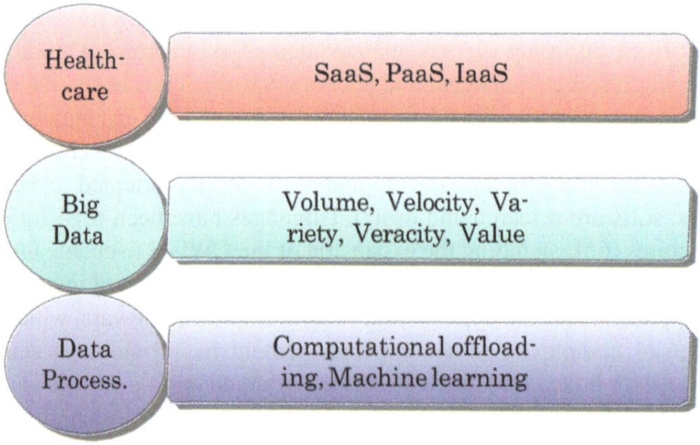

Fig. 4 Role of Could in IoMT

6.2 Big Data Management

We can generally characterize the big data by its 5 V's. Where,

1. Volume— amount of data generated.
2. Velocity—speed at which it produces data.
3. Variety—general variation of data types
4. Veracity—confusion about the types of data can be added later
5. Value—valuable information that can be obtained from the huge information collection.

Various cloud-based technologies for WBAN-based systems are considered in [72, 73]. Such three services discussed above are used in the WBAN systems to build the cloud module. SaaS delivers software applications that allow only authorised persons or organizations to work with health data, while virtualization and database management tools are delivered by SaaS, and the hardware and required infrastructure is delivered by PaaS. The system is mainly used to build and store health records and to allow health practitioners, through virtualization and data management tools, to review the status of their patient as required. The only work that a patient has to do is to set up a profile, set up control over who can gain access to their data, and determines if continuous, on-demand or periodic monitoring is what they want. The benefits of big data management with cloud technology are obvious. It allows nearly unlimited storage space, provides many valuable resources and allows patient and doctor accessibility. This allows patient's greater power over their own healthcare and at the same time helps physicians to offer greater effective services without even having to physically visit their patient in person. In addition, comprehensive data management systems designed to meet all 5 big data characteristics would allow for data mining, machine learning, and other types of deep analysis. By discovering previously identified patterns or trends in patient development over an illness, this may lead to new medical findings.

6.3 Data Processing and Analytics

Out of numerous types of data processing that can be achieved with cloud technology, computational offloading and machine learning are the most important ones. Computational offloading can be of great benefit to the complex sensor nodes like those measuring ECG data, BP (blood pressure), or fall-detection accelerometers. The use of cloud computing in [74] is to process complicated raw data in order to return pertinent results to the patient's sensing device. It is a great idea, as it permits sophisticated sensors to use the computing power of the cloud, while also allowing the patient to easily see their tests and send them to a doctor if required. Computational data-processing offloading is discussed in [75] to evaluate ECG structure and assess whether the shape is consistent with CHF. This test is too difficult to carry out on a wearable computer, and is therefore a prime case of cloud technology's usefulness in data processing. Some work has also been done on comparing machine learning algorithms. Deep neural networks (DNNs) in [76] were contrasted with gradient boosting decision trees (GBDT), logistic regression (LR), and support vector machine (SVM) algorithms for prediction of stroke. With 87.3% accuracy, DNNs performed the best while the gradient boosting decision trees and logistic regression algorithms were close in accuracy of 86.8% and 86.6% respectively. The worst is the SVM, with an accuracy of just 83.9%. This is considered in [72] which indicate that DNNs are well suited for predictive tasks but they will need to run on cloud storage systems due to the complexity of the algorithm. Because of these advantages, in IoT healthcare systems computational offloading is important to make sure that even the most complex physiological parameters can be tracked, allowing the patient to have the highest possible quality of healthcare.

7 Sensors and Wearable Hardware

Medical devices may be described as an instrument, computer, machine, contraceptive, implement, implant, apparatus, reagent for in vitro use or other related article, including a part or accessory intended for use in the diagnosis of disease or other conditions or intended to affect the structure or function of the human or other animal body. The IoMT is a group of several devices connected with each other to form a smart network. This may include smart body sensors, linked surgical instruments and scopes, or networked contact between patient and doctor in the medical field. Devices discussed in [72] which are built in the domain of IoMT are:

1. Wearable Devices: In a smart healthcare system, the most basic component is a network of wireless sensors which automatically and continuously measures physiological signal measurements and performs bounded data processing. Any Vital signs in the body such as variation in heart rate, conductance in skin, blood pressure, rate of respiration, blood glucose, saturation of oxygen, and other health-related signals can be detected and comprehended using apt sensors that could be mounted over clothes or straight on the body. These include wearable devices such as smart watches which allow patients to be monitored accurately, continuously, in real-time. Devices include:

    - Location sensors: monitoring patient's position.
    - Body temperature Measurement Sensors: measures the body temperature.
    - Blood pressure sensors: continuously check patients'blood pressure.
    - Biometric sensors: they are used as a security measure for authorizing access (e.g., thumbprint, hair, hand, etc.).
    - monitoring sensors: continuously check heart rate with the help of ECG and PPG.
    - Respiratory sensors: track and measure respiration of patients.

- Sensors based on activity: tracks regular physical activities such as exercise, sleep etc. using gyroscope sensors.
- Muscle movement motion sensors: check the conditions of the movement of the muscles.
- Trackers: logs all fitness activities.
- Glucose sensors: logs the levels of glucose.
- ECG sensors: checks on the heart's electric and muscular activity.
- Pulse oximeters: measure the pulse rate and oxygen level.
- Accelerating sensors: Record the patient's recovery.
- Drug pumps: administer the drug to the patient's body in a set volume.

2. Implantable Devices: Those instruments work inside the body of the patient. Present instruments shall include:

   - Swallowable video capsule: recreates the digestive tract from the inside of the person.
   - Embedded Cardiac: gathers information and transmits information through a radio channel to neighbouring omnipresent networks.
   - Implantable cardioverter defibrillator (often known as ICD): the ICD is a battery-powered device that is placed beneath the skin to keep track of the heart rhythm.

3. Ambient Devices: These apps detect the ambient environment of the patient to track habits of behaviour, quality of sleep, access to the toilet, etc. and send clinicians warnings when suspicious trends are detected. These devices are supposed to make the rooms smarter and healthier for chronic patients. Ambient sensors will include:

   - Motion sensors: the sensing of people's motions in a place.
   - Room temperature sensors: measure the temperature of the room.
   - Pressure sensors: monitor the quantity of water, ice and air.
   - Door sensors: door sensors i.e. open and close for the Alzheimer's patients or for the avoidance of infections.
   - Vibration sensors: Sense body vibrations of patients on sheets.
   - Daylight Sensors: Measure the ambient light to continuously change the lighting areas in the house.

4. Stationary Devices: This category refers to instruments that are constant with the customer, but not usually. These include: Imaging devices: Generate vivid pictures of the body's interior for scientific research and operation, such as magnetic resonance imaging (MRI) and X-rays [72].

8 Concerns, Challenges and Risks

IoMT-related concerns as stated in [56] can be classified into four key categories, one of them is raised by the IoMT users and is related to the security, privacy, trust and accuracy issues.

1. Security Concerns: Due to the reliance of IoMT devices on the use of open wireless communications, these devices are prone to various wireless/network attacks. In fact, an attacker can eavesdrop and intercept incoming and outgoing data and information due to the lack of security measures that most IoMT devices either suffer from by design, or due to weak security authentication measures that can be easily bypassed by a skilled attacker. Another security issue is the ability to gain unauthorized access, without being detected, due to the inability to detect and prevent such attacks. This would result in gaining an elevated privilege, injecting malicious codes, or infecting devices with malware. Moreover, IoMT devices, when hijacked by terrorists, could be used as a mean for targeted assassination. Moreover, IoMT devices can have a negative effect on the psychological state of patients, since these can potentially scare patients, causing them to suffer from a heart-attack due to being surrounded by machines instead of humans. To mitigate the main IoMT security concerns, protection against passive and active attacks is a must.
2. Privacy Concerns: Passive attacks such as traffic analysis leads to privacy issues since it would be possible to gather and disclose information about patients' identity, in addition to sensitive and confidential information. This is a massive threat for patients since the attacker is capable of identifying persons medical conditions, which can be harmful for the patient. Most of these real-case attacks led to a breach of patients' privacy either through the leakage, or through the disclosure of sensitive information. Privacy is more than ensuring the secrecy of sensitive and private medical information. It also entails the need for anonymity, non-linkability, and non-observability.
   Anonymity: a patient should not be identifiable; passive attacks can see what do, but not what you do.
   Non-Linkability: Items of Interest (IoI) such as subjects, messages, events, actions not to be disclosed by passive attacks.
   Non-Observability: It should not be noticeable whether the message was exchanged between sender/receiver.
3. Trust Concerns: Trust of the users can be breached once their personal information is disclosed. Also, it can be a threat to their lives as their personal medical information will be with the hackers.
4. Accuracy Concerns: Mal-functionality of the device is what leads to Accuracy concerns. These concern related incidents resulted in having more than 1400 patients being partially or permanently injured, where reports of malfunction revealed that more than 8061 malfunctions have occurred within thirteen years (2000–2013). These incidents indicate the lack of accuracy and precision in the

operations being led by medical robots, along with the false diagnosis of patients, and wrong medical prescriptions.

IoMT challenges emerged as soon as the integration of medical devices into IoT systems started. One major challenge is the lack of standardization. In [77], authors discussed in detail the main IoMT challenges. The issue of standardization is essential to having different medical devices operating together, and for vendors to adopt the right security measures to protect them from being hacked. This would lead to higher protection, efficiency, scalability, consistency, and effectiveness [56]. The deployment of IoMT systems into the healthcare domain is associated with a number of risks which are listed as follows:

1. Disclosure of Personal Information can seriously affect patient's medical conditions, as well as the hospital's reputation.
2. Data Falsification can result into having the transmitted data from any medical device altered and modified, which would result into a higher drug dosage or wrong medical description that can lead to further medical complications
3. Whistle-blowers are based on unsatisfied or rogue medical employees leaking medical details and information about the hospital or patients by either being bribed, or part of an organized crime activity, risking patients' privacy and lives.
4. Lack of Training among nurses and doctors can result in risking patients' lives with permanent disabilities or the loss of life.
5. Accuracy is still a debatable issue and is still responsible for inaccuracies in the medical operations conducted by specialized robots. This can also seriously affect patients' lives and lead to disabilities or fatalities.

9 Conclusion

In this paper, we presented an outline of IoT technologies, problems and usage of cloud computing in healthcare. The most important areas of use have been addressed and a variety of study advantages have been identified for usage. The Internet of Medical Things (IoMT) devices have the ability to produce data in large amounts that can then be integrated with AI. Thankfully, current IoT solutions have an interface for collecting information from different devices and can provide a fairly simple way to use IoT data in AI or ML models. We just need to increase the precision and accuracy of the existing models. Given its benefits, IoMT is vulnerable to a range of threats, concerns and problems that are directed specifically at the safety of patients and the security, quality and provision of healthcare services. Hence, it is important to ensure a high degree of reliability, privacy, trust and consistency. A series of research issues have been identified, that are projected to become major trends in research in the coming years. We assume that this review can be helpful to researchers and professionals in the area, allowing them to appreciate the immense possibility of IoT in the medical world.

References

1. Kaare, C., Gabriel, D., James, W.V., Roland, R.: Ageing populations: the challenges ahead. TheLancet **374**(9696): 1196–1208 (2001)
2. Yach, D., Hawkes, C., Hofman, K.J., Gould, C.L.: The global burden of chronic dis eases: overcoming impediments to prevention and control. J. Am. Med. Assoc. **291**(21), 2616–2622 (2004)
3. Arthur, G., Youakim, B., Bertrand, M., Ervin, S.: IoMTs: a review of recent contributions dealing with cyber-physical systems in medicine. IEEE Internet Things J. **5**(5), 3810–3822 (2018)
4. Insup, L., Oleg, S.: Medical cyber physical systems. In: Proceedings of the 47th Design Automation Conference. ACM, pp. 743–748 (2010)
5. Big Data in the Internet of Things (IoT): Key Trends, Opportunities and Market Forecasts, pp. 2015–2020
6. Khan, S.U., Islam, N., Jan, Z., Din, I., Khan, A., et al.: The Detection and Classification of Breast Cancer in Breast Cytology Images as an IoMT Application, vol. 98, pp. 286–296. Elsevier, Amsterdam (2019)
7. Digital Strategies for Big PicHofman KJ Ture: Available: https://www.tcs.com/resources/white_papers/Pages/Internet-of-Things-Medical-Devices.aspx
8. Huang, C.H., Cheng, K.W.: RFID technology combined with IoT application in medical nursing system. Bull. Network. Comput. Syst. Softw. 3(1), 20–24 (2014)
9. Puri, C., Ukil, A., Bandyopadhyay, S. et al.: iCarMa: Inexpensive cardiac arrhythmia management—an IoT healthcare analytics solution. In: Proceedings of the First Workshop on IoT-Enabled Healthcare and Wellness Technologies and Systems, pp. 3–8 (2016)
10. Kumar, D.D., Venkateswarlu, P.: Secured smart healthcare monitoring system based on IOT. Imper. J. Interdisc. Res. **2**(10), 4561–4598 (2016)
11. Singh, R.: A proposal for mobile e-care health service system using IoT for Indian scenario. J. Netw. Commun. Emerg. Technol. (JNCET) **6**(1) (2016)
12. Iwendi, C., Khan, S., Anajemba, J.H., et al.: Realizing an efficient IoMT-assisted patient diet recommendation system through machine learning model. IEEE Access **8**, 28462–28474 (2020)
13. Haidari, L.A., Brown, S.T., Ferguson, M., et al.: The economic and operational value of using drones to transport vaccines. Vaccine **34**, 4062–4067 (2016)
14. Pulver, A., Wei, R., Mann, C.: Locating AED enabled medical drones to enhance cardiac arrest response times. Prehosp. Emerg. Care **20**(3), 378–389 (2016)
15. Li, M., Taylor, R.H.: Spatial motion constraints in medical robot using virtual fixtures generated by anatomy. In IEEE International Conference on Robotics and Automation. Proceedings. ICRA'04, vol. 2, pp. 1270–1275 (2004)
16. Jiang, F., Jiang, Y., Zhi, H., et al.: Artificial intelligence in healthcare: past, present and future. Stroke Vasc. Urol. **2**(4), 230–243 (2017)
17. Hunter, J.: Adopting AI is essential for a sustainable pharma industry. Drug Discov. World. 69–71 (2016)
18. Shah, P., Kendall, F., Khozin, S., et al.: Artificial intelligence and machine learning in clinical development: a translational perspective. NPJ Dig. Med. **2**(69), 1–5 (2019)
19. Agrawal, P.: Artificial intelligence in drug discovery and development. J Pharmacovigil **6** (2018)
20. Kiourti, A., Psathas, K.A., Nikita, K.S.: Implantable and ingestible medical devices with wireless telemetry functionalities: a review of current status and challenges. Bioelectromagnetics **35**(1), 1–15 (2014)
21. Marshall, D.R.: Swallowable data recorder capsule medical device. US Patent 6,632,175 (2003)
22. Bandy, W.R., Jamieson, B.G., Powell, K.J., Salsman, K.E., Schober, R.C., Weitzner, J., Arneson, M.R.: Ingestible endoscopic optical scanning device. US Patent 8,529,441 (2013)
23. Liu, X., Steiger, C., Lin, S., Parada, G.A., Liu, J., Chan, H.F., Yuk, H., Phan, N.V., Collins, J., Tamang, S., et al.: Ingestible hydrogel device. Nat. Commun. 10 (2019)

24. Toohey, S.L., Wray, A., Wiechmann, W., et al.: Ten tips for engaging the millennial learner and moving an emergency medicine residency curriculum into the 21st century. Western J. Emerg. Med. **17**(3), 337–343 (2016)
25. Santoso, F., Redmond, S.J.: Indoor location-aware medical systems for smart homecare and telehealth monitoring: state-of-the-art. Physiol. Measur. **36**(10) (2015)
26. McFarland, S., Coufopolous, A., Lycett, D.: The effect of telehealth versus usual care for home-care patients with long-term conditions: a systematic review, meta-analysis and qualitative synthesis. J. Telemed. Telecare 1–19 (2019)
27. Cullin, P., Bergdahl, T.: A telecare system. US Patent App. 16/310,127 (2019)
28. Saeed, N., Manzoor, M., Khosravi, P.: An exploration of usability issues in telecare monitoring systems and possible solutions: a systematic literature review. Disabil. Rehabil. Assistive Technol. 1–11 (2019)
29. Anderson, K., Burford, O., Emmerton, L.: Mobile health apps to facilitate self-care: a qualitative study of user experiences. PLoS ONE **11**(5), 156–164 (2016)
30. Kang, M., Park, E., Cho, B.H., Lee, K.S.: Recent patient health monitoring platforms incorporating internet of things-enabled smart devices. Int. Neurourol. J. **22**(2), 76–86 (2018)
31. Chrysant, G.S.: Peripheral vascular disease is associated with increased pulse wave velocity and augmentation index: clinical implications. J. Clin. Hypertens. (Greenwich). **16**, 788–789 (2014)
32. Mammas, C.S., Geropoulos, S., Markou, G., et al.: Mobile telemedicine systems in the multidisciplinary approach of diabetes management: the remote prevention of diabetes complications. Stud. Health Technol. Inform. **202**, 307–310 (2014)
33. Darkins, A., Ryan, P., Kobb, R., et al.: Care coordination/home telehealth: The systematic implementation of health informatics, home telehealth, and disease management to support the care of veteran patients with chronic conditions. Telemed. e-Health **14**(10), 1118–1126 (2008)
34. Mamta, P.S.: An overview of healthcare perspective based security issues in wireless sensor networks. In: 3rd International Conference on Computing for Sustainable Global Development (INDIACom), pp. 870–875 (2016)
35. Jing, Q., Vasilakos, A.V., Wan, J., et al.: Security of the Internet of Things: perspectives and challenges. Wirel. Netw. **20**(8), 2481–2501 (2014)
36. Al-Fuqaha, A., Guizani, M., Mohammadi, M., et al.: Internet of Things: a survey on enabling technologies, protocols, and applications. IEEE Commun. Surv. Tutor. **17**(4), 2347–2376 (2015)
37. Alsubaei, F., Abuhussein, A., Shiva, S.: Security and privacy in the internet of medical things: taxonomy and risk assessment. In: 2017 IEEE 42nd Conference on Local Computer Networks Workshops (LCN Workshops), pp. 112–120 (2017)
38. Zhang, M., Raghunathan, M., Jha, N.K.: Trustworthiness of medical devices and body area networks. Proc. IEEE **102**(8), 1174–1188 (2014)
39. Khandelwal, S.: This $10 device can clone RFID-equipped access cards easily. The Hacker News (2015). Available: https://thehackernews.com/2015/07/hacking-rfid-access-card.html
40. Anthony, S.: USB Killer, yours for $50, lets you easily fry almost every device. Ars Technica (2016). Available:https://arstechnica.com/gadgets/2016/12/usb-killer-fries-devices/. [Cited: 09-Mar-2020]
41. Hackers can remotely kill hospital patients with drug pumps, IT expert Discovers, RT International. Available: https://www.rt.com/news/266491-drug-pumps-hacking-hospitals/. [Cited: 11-Mar-2020].
42. Jafarnia-Jahromi, A., Broumandan, A., Nielsen, J., et al.: GPS vulnerability to spoofing threats and a review of antispoofing techniques. Int. J. Navig. Observ. 1–17 (2012)
43. Gibbs, S.: SS7 hack explained: what can you do about it? (2016). Available: https://www.theguardian.com/technology/2016/apr/19/ss7-hack-explained-mobile-phone-vulnerability-snooping-texts-calls [Cited: 15-Jan-2020]
44. Multiple Vulnerabilities in Animas. Rapid7 Community And Blog. Available: https://community.rapid7.com/community/infosec/blog/2016/10/04/r7-2016-07-multiplevulnerabilities-in-animas-onetouch-ping-insulin-pump. [Cited: 26-Mar-2020]

45. Kumar, P., Lee, H.J.: Security issues in healthcare applications using wireless medical sensor networks: a survey. Sensors **12**(1), 55–91 (2011)
46. Wright, J.: Five wireless threats you may not know. Available: https://www.sans.edu/cyber-research/security-laboratory/article/wireless-security-1 [Cited: 12-March-2020]
47. Kassner, M.: No surprise, IoT devices are insecure. TechRepublic (2014). Available: https://www.techrepublic.com/article/no-surprise-IoT-devices-are-insecure/. [Cited: 10-Mar-2020]
48. Santana-Mancilla, P.C., Anido-Rifón, L.E., Contreras-Castillo, J., et al.: Heuristic evaluation of an IoMT system for remote health monitoring in senior care. Int. J. Environ. Res. Publ. Health **17**, 1–9 (2020)
49. Kumari, K.A., Padmashani, R., Varsha, R., et al.: Securing Internet of Medical Things (IoMT) using private blockchain network. In: Principles of Internet of Things (IoT) Ecosystem: Insight Paradigm, pp. 305–326 (2020)
50. Weisbaum, H.: Cyber attacks and negligence lead to rise in medical data breaches. NBC News (2016). Available: https://www.nbcnews.com/tech/tech-news/cyber-attacks-negligence-lead-rise-medical-data-breaches-n575471. [Cited: 28-Mar-2020]
51. Biddle, S.: Protecting healthcare institutions from DDoS and ransomware attacks (2016). Available: https://www.healthitoutcomes.com/doc/protecting-healthcare-institutions-from-ddos-and-ransomware-attacks-0001. [Cited: 09-Mar-2020]
52. Shan, Y., Kesidis, G., Fleck, D., et al.: Preliminary study of fission defenses against low-volume dos attacks on proxied multi server systems. In: 2017 12th International Conference on Malicious and Unwanted Software (MALWARE), pp. 67–74 (2017)
53. Bogdanoski, M., Suminoski, T., Risteski, A.: Analysis of the syn flood dos attack. Int. J. Comput. Netw. Inf. Secur. **5**(8), 1–11 (2013)
54. Bellare, M., Kohno, T.: Hash function balance and its impact on birthday attacks. In International Conference on the Theory and Applications of Cryptographic Techniques, pp. 401–418 (2004)
55. Jaiswal, S., Gupta, D.: Security requirements for Internet of Things (IoT). In: Proceedings of International Conference on Communication and Networks, Advances in Intelligent Systems and Computing 508 (2017). https://doi.org/10.1007/978-981-10-2750-5_44
56. Jain, A., Singh, T.: Security challenges and solutions of IoT ecosystem. In: Information and communication technology for sustainable development, pp. 259–270 (2020)
57. Oh, S., Park, S.: Task–role-based access control model. Inf. Syst. **28**(6), 533–562 (2003)
58. Ray, I., Toahchoodee, M.: A spatio-temporal role-based access control model. In: IFsIP Annual Conference on Data and Applications Security and Privacy, pp. 211–226 (2007)
59. Cordeiro, L.C., de Lima, F.E.B., Bessa, I.V.: Survey on automated symbolic verification and its application for synthesising cyber-physical systems. IET Cyber-Phys. Syst. Theory Appl. **5**(1), 1–24 (2020)
60. Hatcliff, J., King, A., Lee, I., et al.: Rationale and architecture principles for medical application platforms. In: IEEE/ACM Third International Conference on Cyber-Physical Systems, pp. 3–12. IEEE (2012)
61. Xia, F., Ma, L., Dong, J., et al.: Network QoS management in cyber-physical systems. In: International Conference on Embedded Software and Systems Symposia, pp. 302–307. IEEE (2008)
62. Dillon, T., Potdar, V., Singh, J., et al.: Cyber-physical systems: providing quality of service (QoS) in a heterogeneous systems-of-systems environment. In Proceedings of the 5th IEEE International Conference on Digital Ecosystems and Technologies Conference, pp. 330–335. IEEE (2011)
63. Liu, J., Zhang, L.: QoS modeling for cyber-physical systems using aspect-oriented approach. In Second International Conference on Networking and Distributed Computing, pp. 154–158. IEEE (2011)
64. García-Valls, M., Baldoni, R.: Adaptive middleware design for CPS: considerations on the OS, resource managers, and the network run-time. In: Proceedings of the 14th International Workshop on Adaptive and Reflective Middleware, pp. 1–6 (2015)

65. Lien, S.Y., Cheng, S.M., Shih, S.Y., et al.: Radio resource management for QoS guarantees in cyber-physical systems. IEEE Trans. Parallel Distrib. Syst. **23**(9), 1752–1761 (2012)
66. Zhang, W., Zhang, L.: Designing and modeling cyber physical systems by a service-based approach. In: International Conference on Software Engineering and Service Science, pp. 668–671 (2015)
67. Zhu, W., Zhou, G., Yen, I.L., et al.: A PT-SOA model for CPS/IoT services. In IEEE International Conference on Web Services, pp. 647–654 (2015)
68. Shirvanimoghaddam, M., Shirvanimoghaddam, K., Abolhasani, M.M., et al.: Towards a green and self-powered Internet of Things using piezoelectric energy harvesting. IEEE Access **7**, 94533–94556 (2019)
69. Park, S.O., Do, T.H., Jeong, Y.S.: A dynamic control middleware for cyber physical systems on an IPv6-based global network. Int. J. Commun. Syst. **26**(6), 690–704 (2013)
70. Pan, Z., Hariri, S., Pacheco, J.: Context aware intrusion detection for building automation systems. Comput. Secur. **85**, 181–201 (2019)
71. Tawalbeh, L.A., Mehmood, R., Benkhelifa, E., et al.: Mobile cloud computing model and big data analysis for healthcare applications. IEEE Access **4**, 6171–6180 (2016)
72. Baker, S., Xiang, W., Atkinson, I.: Internet of Things for smart healthcare: technologies, challenges, and opportunities. IEEE Access **5**, 26526–26530 (2017)
73. Hang, L., Jin, W., Yoon, H., et al.: Design and implementation of a sensor-cloud platform for physical sensor management on CoT environments. Electronics **7**(8), 140–150 (2018)
74. Lin, K., Xia, F., Wang, W., et al.: System design for big data application in emotion-aware health care. IEEE Access **4**, 6901–6909 (2016)
75. Abawajy, J.H., Hassan, M.M.: Federated internet of things and cloud computing pervasive patient health monitoring system. IEEE Commun. Manag. **55**(1), 48–53 (2017)
76. Krishnamoorthy, M., Suresh, S., Alagappan, S.: Deep learning techniques and optimization strategies in big data analytics: automated transfer learning of convolutional neural networks using ENAS algorithm. In: Deep Learning Techniques and Optimization Strategies in Big Data Analytics, pp. 142–153 (2020)
77. Hassanalieragh, M., Page, A., Soyata, T., et al.: Health monitoring and management using internet-of-things (IoT) sensing with cloud-based processing: opportunities and challenges. In: 2015 IEEE International Conference on Services Computing (SCC), pp. 285–292 (2015)

IoT Enabled Technology in Secured Healthcare: Applications, Challenges and Future Directions

Sukriti Goyal, Nikhil Sharma, Bharat Bhushan, Achyut Shankar, and Martin Sagayam

Abstract In current years, Internet of Things has come a long way and is well emerged within many organizations and fields covering the sector of healthcare. The continuous execution of IoT within the field of healthcare will direct to a fast rise in productivity and examination of data. In reference to medical gadgets, developments in technology will enhance the results of patients with superior analytics. Thus, this chapter introduces Internet of Health Things with wearable healthcare systems in detail and shows the inter-association of interaction allowed medical gadgets and their combination to broader scale networks of healthcare to enhance the health of patients, and because of this sensitive behavior of systems related to health. Still, It meets various issues, specifically in regards of security, privacy and scalability. This work also presents an overview of approaches based on IoT for healthcare and healthcare aided living as well. Moreover, this chapter illustrates IoT networks for healthcare and the different characteristics of IoT confidentiality and safety including security needs. Also, how distinct technologies such as augmented reality, big data, cloud computing and many more can be implemented in the reference of healthcare are described and ultimately, some pathways for forthcoming work on healthcare based on IoT based on a collection of challenges and open issues are presented.

S. Goyal · N. Sharma · B. Bhushan (✉)
HMR Institute of Technology and Management, Delhi, India
e-mail: bharat_bhushan1989@yahoo.com

S. Goyal
e-mail: sukriti.s9900@gmail.com

N. Sharma
e-mail: nikhilsharma1694@gmail.com

A. Shankar
ASET, Amity University, Uttar Pradesh, Noida, India
e-mail: ashankar2711@gmail.com

M. Sagayam
Karunya Institute of Technology and Sciences, Coimbatore, India
e-mail: martinsagayam.k@gmail.com

A. E. Hassanien et al. (eds.), *Cognitive Internet of Medical Things for Smart Healthcare*, Studies in Systems, Decision and Control 311,
https://doi.org/10.1007/978-3-030-55833-8_2

Keywords Internet of things · Healthcare · Wearable healthcare system · Sensors · Internet of health things · Security · Privacy

1 Introduction

The healthcare field is in a condition of huge desolation. The services of healthcare are more expensive than ever, global population is aging and the counting of chronic diseases are on a growth. What the developers are approaching is a world where general healthcare would become out of reach to most of the humans [1]. While technology can't prevent the people from ageing or remove chronic disorders at one time, but at least it can make healthcare simpler on a pocket in regards of availability and accessibility. A modern technology, known as the Internet of Things (IoT) has a large applicability in various sectors covering the sector of healthcare. The whole application and services of this technology in the field of healthcare is a hope because it permits medical centers to function more competently and patients to gain superior healthcare. There are numerous related works that covered different aspects of Internet of Health Things (IoHT). IoT technology in the field of healthcare is geared towards enhancing the lives of patients and enables a better way of life by using wearable linked gadgets of IoT [2]. The concept of IoHT gadgets as well as smart gadgets of IoT contains huge ability not only for hospitals, but also, for the self-care of people. Patients that may be frequently monitored by IoT gadgets needs detailed tracing of medical reports and this kind of approach uses sensors to collect precise clinical records to provide recommendations and also, use cloud storage to gather and record the clinical details. Due to this, the conventional idea of medical professionals visiting hospitals to check the patient's vital symptoms is replaced by a substitute provisioning a continuous automatic way of data. The target of this concept is to introduce the readers with a detailed evaluation of this concept and IoT sensors applied in gadgets of health monitoring [3]. As, it is found that, in the healthcare sector, IoT technology provides a numerous advantage, including being capable to invigilate patients more closely and using information for analytics. When it occurs to IoT for medical gadget combination, the concentration is moved towards the user end like blood pressure cuffs, glucose meters and other gadgets constructed to store details of patient's crucial symptoms. This allows healthcare professionals to automatically gather data and give recommendations. But, unluckily, medical industries often do not focus on the security threats of linking these gadgets to the Internet [4]. There is a chance that a zero day corrupt in a healthcare gadget can be used to harm or even kill someone without being identified. And it is clear that, IoT gadgets are significant for health applications as they gather crucial medical information of patients, thus, the security of IoHT is prominent for the systems of healthcare because gadgets of IoT are threatened by multiple vulnerabilities of security [5].

The chapter is organized as follows: Sect. 2 describes IoT enabled technology in healthcare with a model for future systems, including topologies, applications and Services of IoHT. Section 3 presents the classes such as ambient assisted living, remote healthcare monitoring, wearable gadgets, solutions for healthcare using smartphones along with different challenges of IoHT. Section 4 explains the security requirements and challenges in IoHT along with various technologies to revolutionize the Services of Healthcare in IoT. Section 5 shows the future research direction in the field of IoHT followed by Conclusion in Sect. 6.

2 IoT Enabled Technology in Healthcare

A necessary part of human's life is "Healthcare". Undoubtedly, an approach is needed to overcome the strain on systems of healthcare needed to supply high-quality observations of the patients who are in danger. IoT technology has been broadly recognized as an efficient approach to reduce the stress on modern systems of healthcare [6]. It stays a comparatively modern area of research, as well as its efficient usage for healthcare, in a field in its prime stage. IoT technology as well as its propriety for modern systems of healthcare is illustrated under this section of chapter. Also, various pioneering researches towards improving systems of IoHT are explored. With the motive of directing the forthcoming growth of these systems [7], a common as well as standardized protocol for forthcoming end-to-end systems of IoHT constructing on the periodic themes are described. This technology has become an industry-agnostic jargon to delineate how the technology has recently been installed in various fields, including the area of healthcare, as well as developing the manner in which business is helmed. The manner in which stakeholders recently communicate with and will continue to streamline the procedures aggregated in frontline works of healthcare has developed by IoT technology. It is a system of physical gadgets that uses connectivity to allow the transmission and swapping of information. But it is not necessary that these gadgets are the complicated for technological developments. However, they perform streamline mechanisms as well as allow staff of healthcare to finish the actions in a timely way. Organizations that are professional in the field of healthcare or technology tend to heavily invest in IoT technology [8]. In existing time period, most of the technological gadgets come with some type of associativity, from wearables namely, Bluetooth or bio-sensors to X-ray systems with Wi-Fi. Those gadgets which are IoT powered, supply sensitive information that help health practitioners execute their tasks [9]. The IoT in the area of healthcare simplifies the crucial operations or jobs to make better patients' results, as well as reduces some of the pressure of health professionals. Jobs namely, treatment improvement observation, remote patient invigilating, and also, the housing of vaccines are all abilities of gadgets of medical combined with the IoT [10].

2.1 Introduction to IoHT

The existing technology in the area of healthcare and a standard practice of medicine gets improved with the systems of IoT. Practitioners reach is extending within a vantage. The different-different data gathered from huge sets of actual world circumstances develops both the size as well as the precision of the medical data. The accuracy of healthcare supply is also enhanced by consolidating more debasing technologies in the networks of healthcare [11]. The employment of IoT in the field of healthcare such as the corporations, personal healthcare and healthcare payment applications, have been heavily grown across several particular use cases of IoT. While, at the same time, other use cases of IoHT are picking up speed and the linked healthcare actuality is speeding up, even if obstacles stay. Also, some of the applications of IoT in the field of healthcare are shown in Fig. 1.

- *Care*: IoT Technology enables practitioners of healthcare to apply their knowledge as well as training in a superior manner to figure out and resolve complications. It supports them to optimize better information and machineries that in turn empowers more accurate and fast tasks. IoT permits in the specialized enhancement of healthcare practitioners because they practically perform their skills rather than contributing time on administrative actions.
- *Gadgets*: Even existing gadgets are developing in their efficiency, accuracy and also, utility, they still supply fewer advantages and features that an IoT system provides. IoT has the capability to unscrew current technologies and direct the humans towards a better healthcare as well as medical gadget solutions. IoT efforts and stows the spaces between the manner we provide healthcare and the machineries by constructing a system instead of creating just weapons. Then, it finds faults and exposes patterns and traceless components in systems of healthcare and recommends developments.

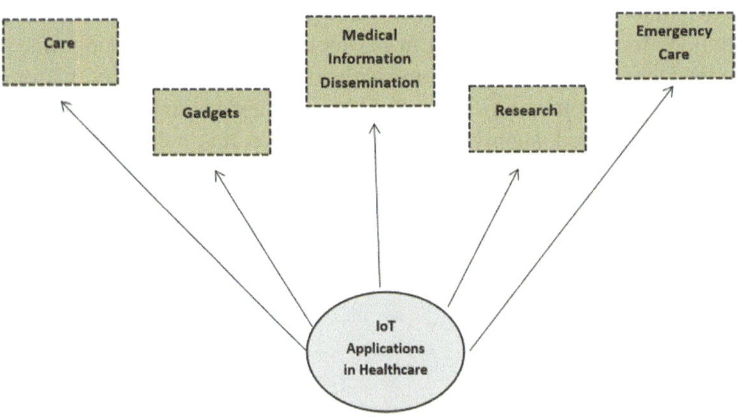

Fig. 1 IoT applications in healthcare

- *Medical Information Dissemination*: The dissemination of precise and recent data to patients' lefts as one of the most defying considerations of medical care, thus, in the field of healthcare, this is one of the most important innovations in the applications of IoT. Gadgets of IoT not only enhance health in the day-to-day lives of humans, but also, support specialist practices. IoT system take care of health out of conveniences such as hospitals and also, permit intrusive care into the home, office or social space. They support as well as allow individuals cater to their own health and permit suppliers of healthcare to provide superior care to patients.
- *Research*: There is a deficiency of sensitive real-world data in the sources that are used by the existing medical research. Most of the time, it uses remaining, monitored surroundings and step forward for medical investigation. Here, IoT technology uncovers paths to an ocean of crucial information and data through explication, real-time field information and also, testing. The data can be provided by the IoT that is far better to general analytics through making usage of equipment's that are able for executing efficient research. As an outcome, IoT supports in the field of healthcare by delivering more practical as well as efficient data which generates superior solutions and disclosure of problems that were unknown in the past, that's why in the area of healthcare, research is one of the most prominent applications of IoT.
- *Emergency Care*: The emergency support facilities have always had the issue of suffering from finite sources and getting disassociated with the root service. In the sector of healthcare, this problem is solved by the enhanced automation as well as analytics of IoT. A case of emergency can be explored from a far distance or rather miles away. The professionals get access to the profiles of patient way before their arrival because of which they can provide necessary care to the patient on time. In this manner, linked losses are decreased, and emergency healthcare is developed.

Apart from these applications, most of the verges of IoT in the sector of healthcare revolved around the advancement of care as such with remote invigilating and telemonitoring as major applications in wider scope of telemedicine. A second area where many verges are present is tracing, invigilating and handling of assets, through RFID as well as IoT [12]. This is implemented on the stage of medical gadgets and assets of healthcare, the people stage and, also the non-medical asset stage for example, hospital constructing assets. But these use cases and deployments are just the starting and, at the same period, are away from omnipresent [13]. More developed and unified techniques within the scope of the digital modification of healthcare are beginning to be used with respect to the health information facets where IoT plays a role of rising approach [14], as it does in particular sectors namely, personal healthcare, smart pills, robotics, smart home care and even, RTHS (real-time health systems).

2.1.1 Applications and Services of IoHT

The healthcare systems based on the technology of IoT can be used in a various number of sectors such as health care for pediatric as well as old aged patients, the inspection of eternal illness, and the supervision of personal fitness and health as well. To understand this concept in detail, the chapter widely groups this section in two facets that are, applications and services where applications are further categorized into two classes: single-condition applications and clustered-condition applications [15]. An application of single-condition implies a particular disease or illness or weakness while on the other side, an application of clustered-condition refers to a number of circumstances or disorders together as a whole. Further, in facilities of IoT, applications of IoT deserve more significance. It can be observed that facilities are used to enhance applications, so that these applications are straightly used by consumers as well as patients. Thus, applications are user-concentrated while, services are developer-concentrated [16]. In this part, several devices, wearable systems, and other healthcare gadgets recently present in the industry are also included. These systems can be considered as IoT innovations that can generate several solutions for healthcare. Figure 2 shows the application of IoHT. This categorization can be simply developed by adding further services with different characteristics and various applications including both single-condition as well as clustered-condition applications.

There is no general description of IoT services in the reference of the healthcare sector. But there may be some conditions in which a service cannot be differentiated from a specific approach. This chapter introduces that an IoHT services is by some mediums common in behavior and has the capability of constructing a section for a collection of approaches or applications [17]. Moreover, it should be considered that common services as well as protocols needed for IoT groundworks may need some alterations for their appropriate processing in healthcare situations. Figure 3 shows different Services of IoHT.

2.1.2 Topologies of IoHT

The IoHT topology implies to the organization of distinct components of an IoHT and implies illustrative situations of seamless surroundings of healthcare. In Table 1, different topologies are presented.

2.2 *A Prototype for Forthcoming Systems of IoHT*

The various needs of the structure of healthcare systems based on IoT become clear after observing the broad variety of such current systems. It is also highlighted by the current systems that interactions are necessary for the network of IoHT [20].

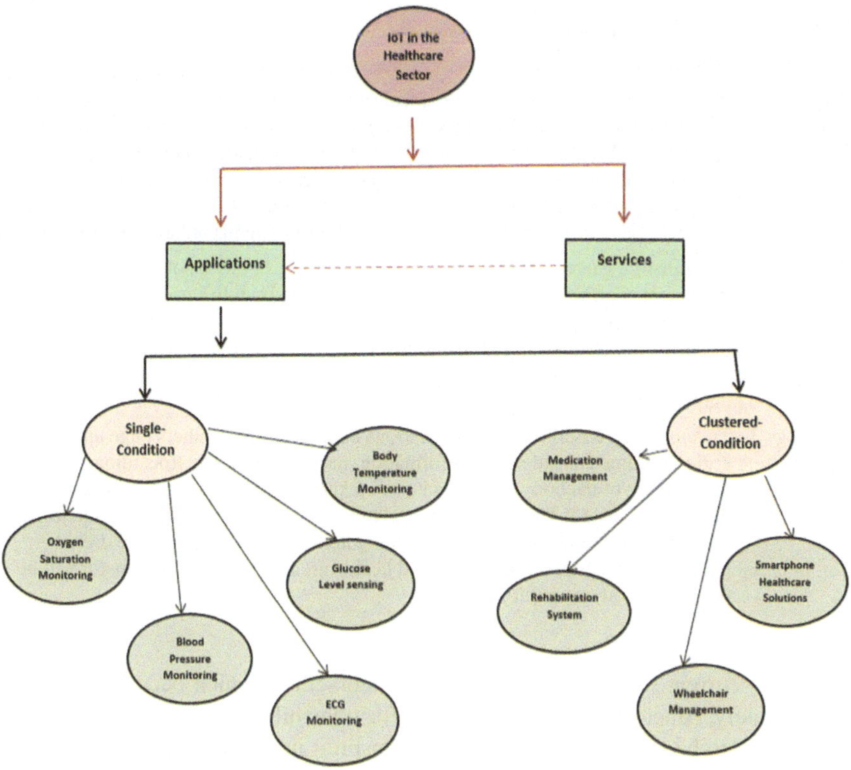

Fig. 2 Applications of IoHT

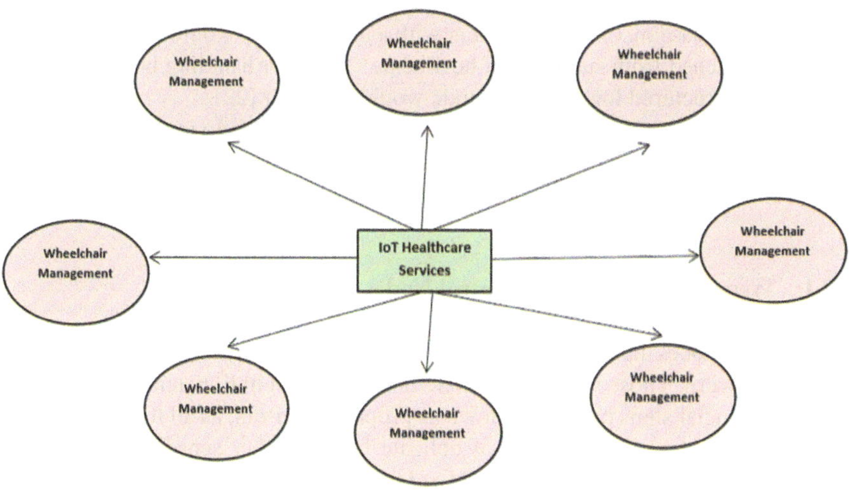

Fig. 3 Services of IoHT

Table 1 Different IoHT topologies [18, 19]

Topologies	Descriptions
Topology 1	It presents a situation in which the medical profile of a patient and important details is collected using portable healthcare gadgets as well as sensors associated with his or her body. Then, collected information is examined and further, recorded as recorded information from several sensors as well as machineries become helpful for collection. According to the examinations and collections, care providers can observe a patient from any place and react according to his or her condition
Topology 2	The role of a gateway is described by this topology. In this topology, iMedPack i.e. intelligent pharmaceutical packaging is used which is an IoT gadget that handles the issue of misusage of medicines by assuring pharmaceutical adherence. The iMedBox also known as intelligent medicine box is implied as a gateway of healthcare with a sequence of several needed sensors and interfaces as well, of many wireless standards
Topology 3	This topology describes how multifarious computing grid gathers huge amounts of critical parameters and sensor information, namely, ECG (electrocardiograms), the temperature of the body, oxygen saturation and BP (blood pressure) and also, generates a basic IoThNet (Internet of Things Network for Healthcare) topology. The different storage as well as computing ability of static and mobile digital gadgets like smartphones, laptops and healthcare terminals is also transformed by it into a hybrid computing grids

In various current network prototypes [21], interactions which occurred in short-range namely, Bluetooth are recommended for transmitting data of the sensor to a smartphone to be executed and in case of long-limit interactions, namely, LTE can be used to transmit the executed data from the user to the practitioner of healthcare, generally, a doctor or surgeon through the Internet or SMS. But there is a major disadvantage of this that basically smartphone has a finite life of battery, needing continuous recharging of it; so, a patient with a finish battery would be considered as a patient detached from suppliers of healthcare. Thus, for handling healthcare data, a particular structured low-powered node would be better [22].

A four-section prototype is described in Fig. 4 that will help in the advancement of forthcoming systems of IoHT. In the coming parts of this section, each of the elements of the recommended prototype is illustrated in detail.

2.2.1 Wearable Sensors and Central Nodes

Those sensor nodes that mete physiological circumstances are known as sensor nodes that can be worn. Those sensors that mete the rate of respiration, temperature of the body, and crucial signs pulse are known as proposed sensors; as all these mentioned signs are necessary for detection of delicate health. Other sensors that could be processed are blood oxygen and also, blood pressure sensors as these signs are often taken side by side the mentioned significant parameters in this point. Apart from these sensors, there are some special-purpose sensors namely, fall detection, joint

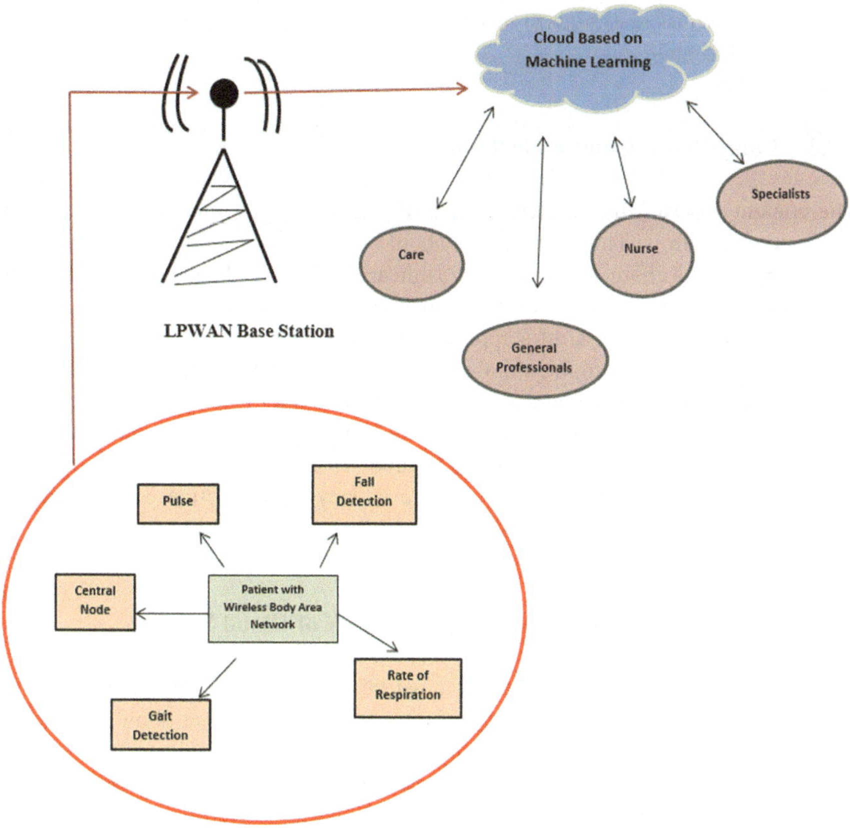

Fig. 4 An overview of the recommended prototype

angle, and blood-glucose sensors which could also be processed for systems aiming a particular circumstance [23].

2.2.2 Short-Range Communications

A short-range interactions approach is needed for sensors to interact with the central node. There are various significant needs to concern when opting a base of interactions of short-limit including effects on the body of a human, delay, and privacy. The opted approach should have no hazardous impacts on the body of a human as any kind of effect could produce more health considerations for patients. The powerful privacy methods should also be provided by it to assure that critical information about the patient cannot be exploited by an attacker. In these types of systems, time latency could be the distinction between the death and life so, low delay would not require to be prioritize as highest in applications that are not time sensitive, however, it is still preferential.

2.2.3 Long-Range Communications

The content received by the central node is not of any use unless something can be processed with it. This information should be transferred further to a database where appropriate characters, namely, healthcare professionals or care providers can safely access it. Again, there are various concerns when opting a relevant long-distance interactions standard for application in healthcare system, including protection of data, low-delay, high availability, fault improving abilities and robustness against any disturbance. As in case of short-distance interactions, strong protection of data is significant to assure that critical data of patient stay confidential and cannot be changed or transformed by anyone. In time-sensitive applications, low-delay is prominent such as urgent healthcare where latency in interaction could have bad impacts on patients.

2.2.4 Machine Learning and Safe Architecture of Cloud Storage

The healthcare received from patients must be safely recorded for further usage. Medical practitioners have advantage from getting information about a patient's medical past, and the technology of machine learning is not efficient unless huge databases of data are present to it. It is found in the literature that cloud storage is the most feasible methodology for recording information. But, giving accessibility for healthcare specialists without conciliating data protection is a major consideration [24, 25] that should be introduced by researchers enhancing systems of IoT healthcare.

2.3 *Wearable Healthcare Systems*

WBANs (Wireless Body Area Networks) have been detected as a major element of a network of healthcare established on the IoT technology, and as such the enhancement of precise sensors with minimum form factor is necessary for the appropriate advancement of this kind of a network. Thus, this section concentrates on sensors that are non-aggressive and non-imperious [20, 26]. Some of the basic sensors are presented in the subsections below and depicted in Fig. 5.

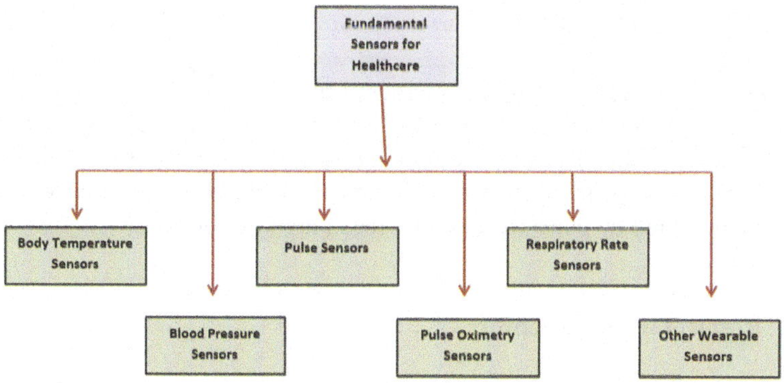

Fig. 5 Fundamental sensors for healthcare

2.3.1 Body Temperature Sensors

Body temperature is an important parameter which can be monitored to find fever, heat stroke, hypothermia, and any other problem. In the context of healthcare, the temperature of a body is a helpful observation weapon that should be constituted in a system that can be worn. The precision of temperature perceiving is finite by how nearly the sensor can be placed to the patient's body.

2.3.2 Blood Pressure Sensors

Whilst not a prominent parameter itself, blood pressure (BP) is constantly meted side by side the other prominent parameters or signs. For cardiovascular disease such as heart attack, high blood pressure (or hypertension) is known as a danger factor. Also, it is one of the most general chronic diseases, affecting 32% of adults. As such, emerging blood pressure into a WBAN for healthcare would give significant data of numerous users [27]. But, still creating a sensor that can be worn for frequently and non-aggressively observation of blood pressure is a challenge in the sector of IoHT.

2.3.3 Pulse Sensors

Pulse is considered as a most important sign which can be used to identify a broad limit of urgent circumstances, namely, pulmonary embolisms, vasovagal syncope and cardiac arrest. Pulse sensors have been broadly examined, both for fitness tracing as well as healthcare reasons. A pulse can be noted from the wrist, fingertip, earlobe, chest and more. High precision is provided by fingertips and earlobe as well, but they are not hugely can be worn. A chest-worn system can be worn, but sensors places on wrist have basically found the most compatible for a long duration system that can be worn [28].

2.3.4 Pulse Oximetry Sensors

The level of oxygen in the blood is measured by pulse oximetry. Similar to blood pressure, level of blood oxygen is not a significant indicator, but, does act as a sign of breathing functions and can support in monitoring of situations like hypoxia. Thus, pulse oximetry is a worthy inclusion in a standard health observation system. Blood oxygen is measured by pulse oximetry by receiving PPG signals.

2.3.5 Respiratory Rate Sensors

Rate of respiration is another prominent indicator. The count of breaths a patient takes per 60 s is known as breathing rate. Examining respiration could help in the detection of problems like apnea episodes, interruptions in the passage of air, asthma attacks, hyperventilation due to panic attacks, lung cancer and many more. Many past works have designed sensors for showing the rate of respiration. Based on thermistor, the first respiratory rate sensor is a nasal sensor [29].

3 Classes and Challenges of IoHT

IoHT is generally a solution based on IoT that constitutes a system structure that permits the link between healthcare services and a patient. The information perceived from patients can be gathered through actuator or sensors and executed by applications produced for a user terminal, namely, smart watches, computers, smart phones or a particular embedded gadget [30]. The terminal of the user is linked to a gateway through short-range interaction protocols, like Bluetooth, BLE (or Bluetooth low energy), or 6LoWPAN (IPv6 over low power wireless personal area networks) over the IEEE 802.15.4 standard [31]. This gateway links to a clinical server or cloud services for execution and storage of data. While on the other side, information of patients can be recorded in a health data system using eHealth records, and when the patient visits a healthcare practitioner, then, he or she can without any problem access the clinic past of the patient.

3.1 Four Basis Classes of IoHT

To understand this topic in detail, this section categorizes the IoHT in four basic classes: ambient assisted living, remote healthcare monitoring, wearable gadgets, and solutions for healthcare using smartphones. All of these are discussed in coming sub-sections.

3.1.1 Ambient Assisted Living

A service based on the IoT technology that assists care of non-viable or elderly patients is known as AAL i.e. ambient assisted living. This kind of solution wants to increase the independent life of the humans in their residences by giving more security. Linking users to smart things, like motion sensors, respiration rate sensors, blood pressure sensors and pulse rate sensors, is a general use of this facility. Not only a securer atmosphere is provided by ambient assisted living, but, also, autonomy is increased and the user is stimulated by AAL to have a longer and more active life [32].

3.1.2 Remote Healthcare Monitoring

Generally, clinics, homecare, and atmosphere of hospitals adopt technologies of remote health monitoring to remotely invigilate the important indicators of a human interacting in real-time to parents, patients and any other individual by reducing the costs of hospital, enhancing quality of care and decreasing the clinician duration. Distant monitoring of health can be executed by the applications that obtain physiological content from the patients to be accessed distantly. User interfaces (such as tablets, computers, laptops, smartphones, and more), a data perceiver (namely, biosensors), and an Internet connection is typically included in these applications. Thus, it is found in [33] that monitoring can be executed with the combination of mobile computing, cloud storage, the architecture of data interaction, and IoT as well. This scheme targets to collect, transfer, record and turn available the visualization of biomedical signs in real-time [34].

3.1.3 Wearable Gadgets

Those smart gadgets that can be connected to the body, such as shoes, body sensors or watches, are known as wearables. These gadgets are capable to link to physiological transmitters to display patients' signals, for example, blood pressure, respiration rate, body temperature, pulse rate, and many others [35]. The use for monitoring patient's physical activities is a general usage of wearable gadgets. Gadgets that can be worn are also applied to provide care to old aged people.

3.1.4 Solutions for Healthcare Using Smartphones

Smartphones are used by many solutions of IoHT. In addition, nowadays, without concerning mobile health aid, ambient aided living solutions cannot be recommended and structured. In this chapter, applications that assist clinical interaction, diagnosis, supplying medicine recommendations or healthcare knowledge and training are considered as solutions for healthcare using smartphones. These solutions target to generate a link between smartphones, sensors, and healthcare practitioners. The field of IoHT is a complicated sector of research showing several issues. The major challenges of the IoHT are introduced in this section.

3.2 Challenges of IoHT

To determine various types of multifarious applications such as traffic control, smart cities, telemedicine or assisted living, embedded networks are used in several surroundings. In these usages, cyber systems also known as digital systems are commanding physical things, which results in a stationary communication between the physical as well as digital world [36]. But, the delicate facets of these applications, specifically when embedded systems are concerned, uplift challenges that can be grouped into three different classes:

- IoHT must depend on precise prototypes of hybrid systems. They available at the scission of both the physical as well as digital world, hence, requiring both precise physical prototypes and accurate computing abstractions. Furthermore, depending on a prototype-based structure allows the enhancement of the testing processes through simulation.
- There must be particular authentication and induration methods: the most of the IoHT is limited to be broadly disseminated and, also, they must include authentication and induration protocols on distinct stages of granularity to pass certifications.
- These systems must fulfill robustness, safety and flexibility needs. Systems based on IoHT are not only capable for preserving admissible performance such as alterations, but also respond accurately if deemed significantly, because of the unsteady behavior of the physiological as well as physical world. Also, some safety considerations are raised by these systems because they often adjust circumstances where system failure could be considered as life threatening. Thus, these networks must be capable to prevent several perpetrator invasions.

4 Security Requirements and Challenges in IoHT

IoT Technology brings revolution in the industry. In future, the healthcare sector is required to hostile the broad disperse acquisition of the IoT technology and develop through modern electronic health gadgets and applications as well. The gadgets and applications of healthcare are required to deal with the special data like confidential information of healthcare [37]. Also, these smart gadgets may be linked to whole world data systems so that, it can be access from anywhere at any time [38]. Thus, the sector of IoHT may be a goal of the perpetrators. It is delicate to find and examine different characteristics of IoT confidentiality and protection, constituting security needs, risks, and threat prototypes from the healthcare point of view, to facilitate the

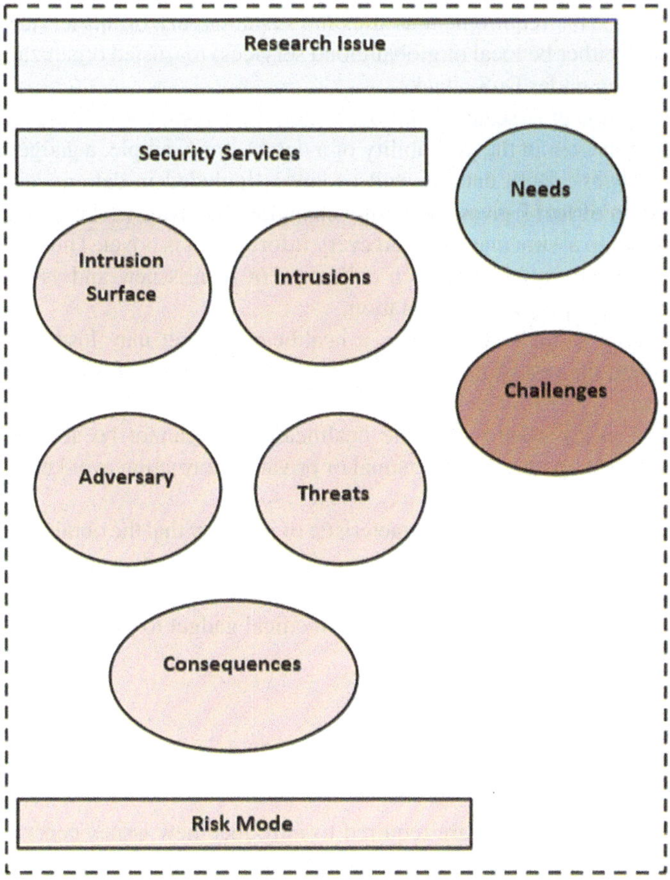

Fig. 6 Security issues in healthcare based on IoT

whole acquisition of the IoT in the field of healthcare. Some of the security issues in healthcare sector are presented in subsections below and depicted in Fig. 6.

4.1 Security Requirements in IoHT

For healthcare solutions based on IoT, security requirements are similar to those general interaction's circumstances. Therefore, the need of security in IoHT systems are explained as follows [39]:

- *Resiliency*: A security approach should still secure the gadget or system or data from any attack if some inter-associated medical gadgets are compromised.

- *Availability*: This requirement assures the sustainability of the services of IoHT (which can either be local or global/cloud services) to trusted organizations when required even under DoS attacks.
- *Defect Tolerance*: A security approach should keep supplying required services of protection even in the availability of a defect, for example, a gadget failure.
- *Data Freshness*: Both data as well as key is included in data novelty because each system of IoHT gives some time changing metering's. Therefore, there is a requirement to assure that each and every information is novel. Thus, information novelty typically refers that each collection of data is new and assures that no adverse party replays old information.
- *Self-Healing*: In an IoHT system, a healthcare gadget may lose or run out of power, then, the rest of the cooperating gadgets should perform a small level of protection.
- *Privacy*: Privacy assures that the healthcare data cannot be accessed by any untrusted participants. Also, personal or private information avoid exposing their data to adversaries.
- *Integrity*: It is assured by the characteristic of integrity that the obtained healthcare contents are not changed during transmission by any adversary. Moreover, the genuineness of gathered content and information should not be compromised.
- *Verification*: This feature allows an IoT medical gadget to assure the detail of the system with which it is interacting.

4.2 Security Challenges in IoHT

The novel counter practices are required to introduce new issues generated by the IoT because needs for IoT protection are not assured by conventional protection strategies [40]. Some challenges for securing IoHT facilities are as follows:

4.2.1 Energy Limitations

A general IoT network for healthcare constitutes small medical gadgets of finite battery power. These gadgets save energy by enabling the power saving mode when no sensor observation requires to be noted. Moreover, they perform at a low speed of CPU if there is nothing crucial to be executed [41]. Thus, the power constraint feature of IoT medical gadgets makes searching an energy aware protection solution is a huge challenge.

4.2.2 Dynamic Security Updates

There is a huge requirement to keep security protocols updated to reduce high risk vulnerabilities. Thus, up-to-date protection patches are required for IoT medical gadgets. But, structuring a method for faster installation of protection systems is a defying job.

4.2.3 Multiplicity of Gadgets

The healthcare gadgets within an IoHT system are different, limiting from full-fledged PCs to low-end RFID tags. These gadgets differ according to their ability in regards of their computing ability, energy, embedded software, and memory [42]. Thus, the issue lies in structuring a protection approach that can accommodate the gadgets.

4.2.4 Scalability

The number of IoT gadgets has risen continuously, and thus, numerous gadgets are getting linked to the whole world data network. Hence, constructing a hugely flexible protection approach without compromising the needs of security becomes a huge challenge.

4.2.5 Memory Limitations

Most of the gadgets of the IoHT systems have minimum on-gadget memory due to which these gadgets are enabled using an integrated OS (operating system), a binary application, and system software. Thus, their memory may not be enough to perform complex protocols of security.

4.2.6 Multi-protocol Network

Through a possessory network protocol, a medical gadget may interact with other gadgets in the local network. Moreover, the similar IoT gadget may interact with facility suppliers of IoT over the IP network. Thus, it is found tough by protection experts to find a better security solution for multiple protocol interactions.

4.2.7 Computational Limitations

IoT medical gadgets are combined with processors having low-speed, due to which the central processing unit, i.e. CPU in these gadgets is not very efficient in regards of their speed [43]. Moreover, these gadgets are mot structured to execute operations, which are computationally expensive means they generally operate as an actuator/sensor [44]. Thus, searching a protection solution that reduces the more usage of sources and hence, increases safety is a defying job [45].

4.3 *Various Technologies to Revolutionize the Services of Healthcare in IoT*

For healthcare solutions based on IoT, there are multiple potential technologies, and thus, it is a vary task to set up an explicit list [46]. Therefore, in reference of this, this section concentrates on various origin technologies that have the capability to revolutionize the services of healthcare that are based on the technology called IoT [47].

- *Ambient Intelligence*: It is prominent technology because clients, consumers, and end users in a healthcare system are human beings such as healthcare providers or patients [48]. The frequent learning of human being nature and execution of any needed action triggered by an acknowledged operation is allowed by ambient intelligence.
- *Augmented Reality*: In this, the characters plays a major role in the sector of healthcare engineering [49]. It is applicable for remote invigilating and surgery.
- *Big Data*: A large quantity of necessary medical information created from different medical sensors and supply equipment's for developing the potential of proper health diagnosis as well as invigilating methodologies and levels as well, are included in big data [50].
- *Cloud Computing*: The combination of cloud computing technology into the healthcare technologies based on IoT should supply services with universal access to exchanged sources [51], providing facilities upon on urge over the system and processing tasks to fulfill several requirements [52].
- *Grid Computing*: By integrating grid computing technology into the universal systems of healthcare [53], the not sufficient computational ability of medical sensor nodes can be focused. The grid computing technology, more precisely cluster computing, can be considered as the base of cloud computing technology [54].
- *Networks*: Several networks limiting from networks for short distance interactions (such as 6LoWPANs, WBANs, WSNs, WPANs and WLANs) [55] to long distance interactions (such as any kind of cellular network) are portions of the physical architecture of the healthcare network based on IoT.
- *Wearables*: Engagement of patients as well as health enhancements of the population can be serviced by accepting wearable medical gadgets as land tags [56]. This has a number of primary advantages: goal-oriented, healthcare centers, and linked data [57].

5 Future Research Direction

Many investigators have researched on constructing as well as executing several healthcare facilities based on the IoT [58], and also, on resolving several structural and technical issued related to those facilities [59]. There are various other open issues and challenges as well, that need to be sharply focused in addition to research considerations in the literature. Some of the challenges and open issues are presented in the subsections below and depicted in Fig. 7.

- *Identification*: The communities or organizations of healthcare typically deal with multiple patient surroundings in which many care providers fulfill their jobs [60]. From this point of view, the most suitable and appropriate recognizance of patients and care providers as well, is essential [61].
- *Cost Analysis*: It may be perceived by investigators that healthcare facilities based on IoT is a low-cost technology [62], but it is observed that, no evidence has provided by the study. Thus, in this reference, a cost-analysis of a general IoHT network may be useful.

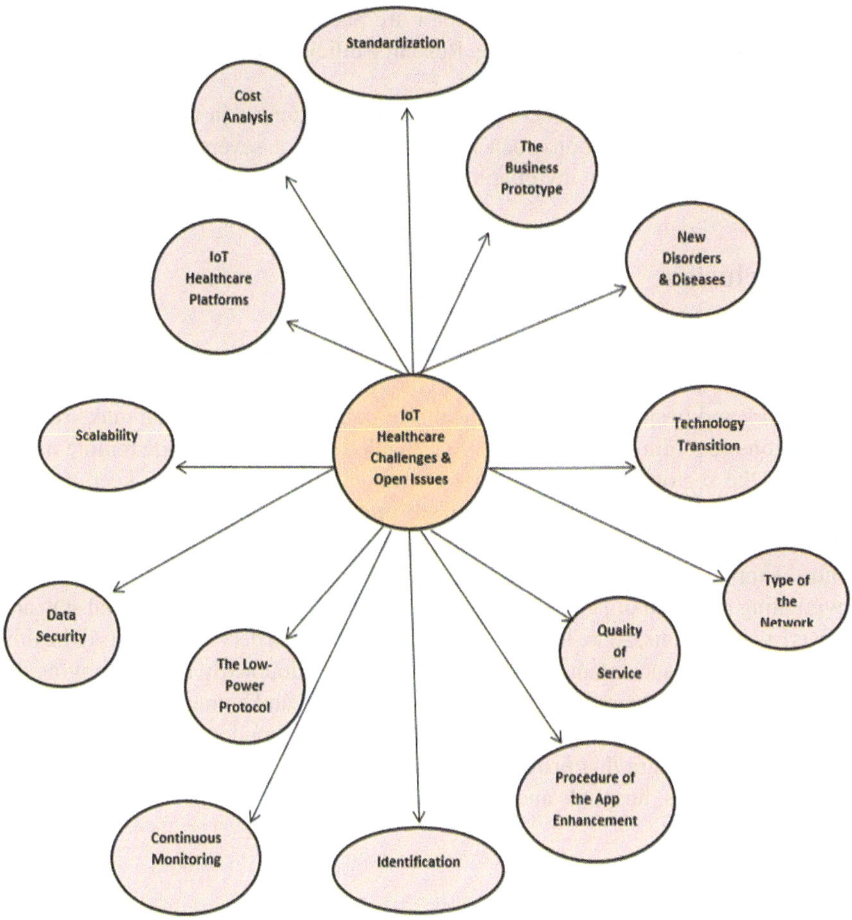

Fig. 7 Challenges and Open Issues Related to IoHT

- *Business Prototype*: This approach of IoHT is not yet robust because it includes a collection of components with modern needs, namely, modern operational procedures as well as policies, modern systems of architecture, disseminated targeted consumers, and changed organizational designs [63]. Moreover, healthcare professionals and nurses typically neglect learning and applying modern technologies. Thus, there is an imperious requirement for a modern and efficient business prototype.
- *Data Security*: The security of collected medical information from several sensors and gadgets from illegal access is important. Thus, technological protection practices and policies should be familiarized to exchange medical information with trusted users, applications, and organizations [64]. In this, familiarizing an efficient algorithm for complicity among security, recognition and respond facilities to avoid several intrusions, risks and vulnerabilities is a kind of open issue. Here, various challenges related to this sector are Security of managing IoT, big data, Physical security, Security routing, Resource efficient security, Data transparency [65].
- *Continuous Monitoring*: There are a lot of circumstances in which patients need long duration observation, such as a patient with a severe disease. Thus, the requirement of frequent invigilating as well as logging is crucial in this reference.

6 Conclusion

Before IoT technology, communications of patients with medical professionals were finite to visit, tele-interactions, and text communications. There was no other way for doctors and hospitals to monitor health of patients frequently and make recommendations according to their health. IoT enabled gadgets have made remote monitoring in the sector of healthcare possible, unleashing the ability to keep patients healthy and safe, and also helping medical professionals to provide better care. It has also improved satisfaction as well as engagement of patients as communications with doctors have become simpler and more efficient in nature. Moreover, distant invigilating of health of patients helps in decreasing the duration of hospital stay and resists re-admissions. Also, IoT has a primary influence on decreasing costs of healthcare prominently and enhancing treatment results. Undoubtedly IoT is changing the sector of healthcare by redefining the space of gadgets and human's communication by providing healthcare solutions. IoT has a numerous applications and services in the field of healthcare that provide advantage medical professionals, patients, care providers, families, hospitals and even insurance companies.

References

1. Mutlag, A.A., Ghani, M.K., Arunkumar, N., Mohammed, M.A., Mohd, O.: Enabling technologies for fog computing in healthcare IoT systems. Fut. Gener. Comput. Syst. **90**, 62–78 (2019)
2. Min, M., Wan, X., Xiao, L., Chen, Y.B., Xia, M., Wu, D., Dai, H.: Learning-based privacy-aware offloading for healthcare IoT with energy harvesting. IEEE Internet Things J. **6**, 4307–4316 (2019)
3. S.R. Moosavi, E. Nigussie, M. Levorato, S. Virtanen, J. Isoaho: Performance Analysis of End-to-End Security Schemes in Healthcare IoT. ANT/SEIT (2018)
4. Gope, P., Hwang, T.: BSN-Care: a secure IoT-based modern healthcare system using body sensor network. IEEE Sens. J. **16**, 1368–1376 (2016)
5. Yeh, K.: A secure IoT-based healthcare system with body sensor networks. IEEE Access **4**, 10288–10299 (2016)
6. Mahmoud, M.M., Rodrigues, J.J., Ahmed, S.H., Shah, S.C., Al-Muhtadi, J., Korotaev, V., Albuquerque, V.H.: Enabling technologies on cloud of things for smart healthcare. IEEE Access **6**, 31950–31967 (2018)
7. Zhu, N., Diethe, T., Camplani, M., Tao, L., Burrows, A., Twomey, N., Kaleshi, D., Mirmehdi, M., Flach, P.A., Craddock, I.: Bridging e-Health and the Internet of Things: The SPHERE project. IEEE Intell. Syst. **30**, 39–46 (2015)
8. Henze, M., Hermerschmidt, L., Kerpen, D., Häußling, R., Rumpe, B., Wehrle, K.: A comprehensive approach to privacy in the cloud-based Internet of Things. Fut. Gener. Comput. Syst. **56**, 701–718 (2016)
9. Baker, S.E., Xiang, W., Atkinson, I.M.: Internet of Things for smart healthcare: technologies, challenges, and opportunities. IEEE Access **5**, 26521–26544 (2017)
10. Marques, G., Pitarma, R., Garcia, N.M., Pombo, N.: Internet of Things architectures, technologies, applications, challenges, and future directions for enhanced living environments and healthcare systems: a review. Electronics **8**, 1081 (2019)
11. Elsts, A., Fafoutis, X., Woznowski, P., Tonkin, E., Oikonomou, G., Piechocki, R.J., Craddock, I.: Enabling healthcare in smart homes: the SPHERE IoT network infrastructure. IEEE Commun. Mag. **56**, 164–170 (2018)
12. Chang, S., Chiang, R., Wu, S., Chang, W.: A context-aware, interactive M-Health system for diabetics. IT Profess. **18**, 14–22 (2016)
13. Pasluosta, C.F., Gassner, H., Winkler, J., Klucken, J., Eskofier, B.M.: An emerging era in the management of Parkinson's disease: wearable technologies and the Internet of Things. IEEE J. Biomed. Health Inf. **19**, 1873–1881 (2015)
14. Zhou, W., Jia, Y., Peng, A., Zhang, Y., Liu, P.: The effect of IoT new features on security and privacy: new threats, existing solutions, and challenges yet to be solved. IEEE Internet Things J. **6**, 1606–1616 (2019)
15. Marques, G., Pitarma, R.: mHealth: indoor environmental quality measuring system for enhanced health and well-being based on internet of things. J. Sens. Actuat. Netw. **8**, 43 (2019)
16. Bisio, I., Garibotto, C., Lavagetto, F., Sciarrone, A.: When eHealth meets IoT: a smart wireless system for post-stroke home rehabilitation. IEEE Wirel. Commun. **26**, 24–29 (2019)
17. Wolgast, G., Ehrenborg, C., Israelsson, A., Helander, J., Johansson, E., Månefjord, H.: Wireless body area network for heart attack detection [Education Corner]. IEEE Antennas Propag. Mag. **58**, 84–92 (2016)
18. Zhou, J., Cao, Z., Dong, X., Vasilakos, A.V.: Security and privacy for cloud-based IoT: challenges. IEEE Commun. Mag. **55**, 26–33 (2017)
19. Milici, S., Lorenzo, J., Lázaro, A., Villarino, R.M., Girbau, D.: Wireless breathing sensor based on wearable modulated frequency selective surface. IEEE Sens. J. **17**, 1285–1292 (2017)
20. Kaur, R., Kaur, K., Khamparia, A., Anand, D.: An Improved and adaptive approach in ANFIS to predict knee diseases. Int. J. Healthcare Inf. Syst. Inf. (IJHISI) **15**(2), 22–37 (2020)
21. Costa, S.E., Rodrigues, J.J., Silva, B.M., Isento, J.N., Corchado, J.M.: Integration of wearable solutions in aal environments with mobility support. J. Med. Syst. **39**, 1–8 (2015)

22. Brito, M.S., Hoque, S., Steinke, R., Willner, A., Magedanz, T.: Application of the Fog computing paradigm to Smart Factories and cyber-physical systems. Trans. Emerg. Telecommun. Technol. **29** (2018)
23. Yang, G., Deng, J., Pang, G., Zhang, H., Li, J., Deng, B., Pang, Z., Xu, J., Jiang, M., Liljeberg, P., Xie, H., Yang, H.: An IoT-enabled stroke rehabilitation system based on smart wearable armband and machine learning. IEEE J. Transl. Eng. Health Med. **6**, 1–10 (2018)
24. Khamparia, A., Saini, G., Gupta, D., Khanna, A., Tiwari, S., de Albuquerque, V.H.C.: Seasonal crops disease prediction and classification using deep convolutional encoder network. Circuits Syst. Signal Process. (2019). https://doi.org/10.1007/s00034-019-01041-0.
25. Lee, E.A.: Cyber physical systems: design challenges. In: 2008 11th IEEE International Symposium on Object and Component-Oriented Real-Time Distributed Computing (ISORC), pp. 363–369 (2008)
26. Olaronke, I., Oluwaseun, O.: Big data in healthcare: prospects, challenges and resolutions. In: 2016 Future Technologies Conference (FTC), pp. 1152–1157 (2016)
27. Khamparia, A., Gupta, D., de Albuquerque, V.H.C., Sangaiah, A.K., Jhaveri, R.H.: Internet of health things driven deep learning system for detection and classification of cervical cells using transfer learning. J. Supercomput. (2020). https://doi.org/10.1007/s11227-020-03159-4
28. Islam, S.M., Kwak, D., Kabir, M.H., Hossain, M., Kwak, K.: The internet of things for health care: a comprehensive survey. IEEE Access **3**, 678–708 (2015)
29. Zanero, S.: When cyber got real: challenges in securing cyber-physical systems. IEEE Sens. **2018**, 1–4 (2018)
30. Silva, B.M., Rodrigues, J.J., Díez, I.D., Coronado, M.L., Saleem, K.: Mobile-health: a review of current state in 2015. J. Biomed. Inform. **56**, 265–272 (2015)
31. Habibzadeh, H., Dinesh, K., Shishvan, O.R., Boggio-Dandry, A., Sharma, G., Soyata, T.: A survey of healthcare internet of things (HIoT): a clinical perspective. IEEE Internet Things J. **7**, 53–71 (2020)
32. Machado, F.M., Koehler, I.M., Ferreira, M.S., & Sovierzoski, M.A.: An mHealth remote monitor system approach applied to MCC using ECG signal in an android application. WorldCIST **19** (2016)
33. Singh, A., Mehta, J.C., Anand, D., Nath, P., Pandey, B., Khamparia, A.: An intelligent hybrid approach for hepatitis disease diagnosis: combining enhanced k-means clustering and improved ensemble learning. Exp. Syst. (2020). https://doi.org/10.1111/exsy.12526
34. Rodrigues, J.J., Segundo, D.B., Junqueira, H.A., Sabino, M.H., Prince, R.M., Al-Muhtadi, J., Albuquerque, V.H.: Enabling technologies for the internet of health things. IEEE Access **6**, 13129–13141 (2018)
35. Zinner, T., Wamser, F., Leopold, H., Dobre, C., Mavromoustakis, C.X., Garcia, N.M.: Matching requirements for ambient assisted living and enhanced living environments with networking technologies (Chap. 5) (2017)
36. Tiwari, R., Sharma, N., Kaushik, I., Tiwari, A., Bhushan, B.: Evolution of IoT and data analytics using deep learning. In: 2019 international conference on computing, communication, and intelligent systems (ICCCIS) (2019). https://doi.org/10.1109/icccis48478.2019.8974481
37. Alsamhi, S.H., Ma, O., Ansari, M., Almalki, F.A.: Survey on collaborative smart drones and internet of things for improving smartness of smart cities. IEEE Access **7**, 128125–128152 (2019)
38. Jadon, S., Choudhary, A., Saini, H., Dua, U., Sharma, N., Kaushik, I.: Comfy smart home using IoT. SSRN Electron. J. (2020). https://doi.org/10.2139/ssrn.3565908
39. Rajkumar, R.Y., Lee, I., Sha, L., Stankovic, J.A.: Cyber-physical systems: the next computing revolution. In: *Design Automation Conference* (2010), pp. 731–736
40. Rubí, J.N., Gondim, P.R.: IoMT platform for pervasive healthcare data aggregation, processing, and sharing based on OneM2M and OpenEHR. Sensors 19 (2019)
41. Varshney, T., Sharma, N., Kaushik, I., Bhushan, B. (2019). Architectural model of security threats & their countermeasures in IoT. In: 2019 International Conference on Computing, Communication, and Intelligent Systems (ICCCIS*)*. https://doi.org/10.1109/icccis48478.2019.8974544

42. Yassein, M.B., Hmeidi, I., Alharbi, M.M., Mrayan, L., Mardini, W., Khamayseh, Y.: IoT-based healthcare systems: a survey. DATA '19 (2018)
43. Khamparia, A., Singh, A., Anand, D., Gupta, D., Khanna, A., Arun Kumar, N., Tan, J.: A novel deep learning based multi-model ensemble methods for prediction of neuromuscular disorders. Neural Comput. Appl. (2018). https://doi.org/10.1007/s00521-018-3896-0
44. Jaidka, H., Sharma, N., Singh, R.: Evolution of IoT to IIoT: Applications & challenges. SSRN Electron. J. (2020). https://doi.org/10.2139/ssrn.3603739
45. Arora, A., Kaur, A., Bhushan, B., & Saini, H.: Security concerns and future trends of Internet of Things. In: *2nd International Conference on Intelligent Computing, Instrumentation and Control Technologies (ICICICT)* (2019). https://doi.org/10.1109/icicict46008.2019.8993222
46. Jara, A.J., Zamora-Izquierdo, M.A., Gómez-Skarmeta, A.F.: Interconnection framework for mHealth and remote monitoring based on the Internet of Things. IEEE J. Sel. Areas Commun. **31**, 47–65 (2013)
47. Pang, Z., Chen, Q., Tian, J., Zheng, L., Dubrova, E.: Ecosystem analysis in the design of open platform-based in-home healthcare terminals towards the Internet-of-Things. In: *Proceedings of International Conference on Advanced Communications Technology* (ICACT) (2013), pp. 529–534
48. Fan, Y.J., Yin, Y.H., Xu, L., Zeng, Y., Wu, F.: IoT-based smart rehabilitation system. IEEE Trans. Industr. Inf. **10**, 1568–1577 (2014)
49. Jia, X., Chen, H., Qi, F.: Technical models and key technologies of E-Health Monitoring. In: 2012 IEEE 14th International Conference on e-Health Networking, Applications and Services (Healthcom), pp. 23–26 (2012)
50. Chouhan, V., Singh, S.K., Khamparia, A., Gupta, D., Tiwari, P., Moreira, C., Damaševičius, R., de Albuquerque, V.H.C.: A novel transfer learning based approach for pneumonia detection in chest X-ray images. Appl. Sci. **10**(2), 559 (2020)
51. Aranki, D., Kurillo, G., Yan, P., Liebovitz, D.M., Bajcsy, R.: Real-time tele-monitoring of patients with chronic heart-failure using a smartphone: lessons learned. IEEE Trans. Affect. Comput. **7**, 206–219 (2016)
52. Talal, M., Zaidan, A.A., Zaidan, B.B., Albahri, A.S., Alamoodi, A.H., Albahri, O.S., Alsalem, M.A., Lim, C.K., Tan, K.L., Shir, W.L., Mohammed, K.I.: Smart home-based IoT for real-time and secure remote health monitoring of triage and priority system using body sensors: multi-driven systematic review. J. Med. Syst. **43**, 1–34 (2019)
53. Dash, S., Shakyawar, S.K., Sharma, M., Kaushik, S.: Big data in healthcare: management, analysis and future prospects. J. Big Data **6**, 1–25 (2019)
54. Ramya, S., Karthikeyan, P., Kaviya, K., Ananthi, B., Sachin, M.C.: IoT based smart health monitoring system: a recent approach and analysis. Int. J. Adv. Res. Innov. Ideas Educ. **5**, 1403–1410 (2019)
55. Harjani, M., Grover, M., Sharma, N., Kaushik, I.: Analysis of various machine learning algorithm for cardiac pulse prediction. In: *International Conference on Computing, Communication, and Intelligent Systems (ICCCIS)* (2019). https://doi.org/10.1109/icccis48478.2019.8974519
56. Khamparia, A., Saini, G., Pandey, B., Tiwari, S., Gupta, D., Khanna, A.: KDSAE: chronic kidney disease classification with multimedia data learning using deep stacked autoencoder network. Multimedia Tools Appl. https://doi.org/10.1007/s11042-019-07839-z (2019)
57. Al-khafajiy, M., Webster, L., Baker, T., Waraich, A.: Towards fog driven IoT healthcare: challenges and framework of fog computing in healthcare. *ICFNDS '18* (2018).
58. Goel, A.K., Rose, A., Gaur, J., Bhushan, B.: Attacks, countermeasures and security paradigms in IoT. In: *2nd International Conference on Intelligent Computing, Instrumentation and Control Technologies (ICICICT)* (2019). https://doi.org/10.1109/icicict46008.2019.8993338
59. Farahani, B., Firouzi, F., Chang, V.I., Badaroglu, M., Constant, N., Mankodiya, K.: Towards fog-driven IoT eHealth: promises and challenges of IoT in medicine and healthcare. Fut. Gener. Comput. Syst. **78**, 659–676 (2018)
60. Miotto, R., Wang, F., Wang, S., Jiang, X., Dudley, J.: Deep learning for healthcare: review, opportunities and challenges. Brief. Bioinform. **19**(6), 1236–1246 (2018)

61. Siyal, A.A., Junejo, A.Z., Zawish, M., Ahmed, K., Khalil, A., Soursou, G.: Applications of blockchain technology in medicine and healthcare: challenges and future perspectives. Cryptography **3**, 3 (2019)
62. Tuli, S., Tuli, S., Wander, G., Wander, P., Gill, S.S., Dustdar, S., Sakellariou, R., Rana, O.: Next generation technologies for smart healthcare: challenges, vision, model, trends and future directions. Internet Technol. Lett. 3 (2020)
63. Elhoseny, M., Ramírez-González, G., Abu-Elnasr, O.M., Shawkat, S.A., Arunkumar, N., Farouk, A.: Secure medical data transmission model for iot-based healthcare systems. IEEE Access **6**, 20596–20608 (2018)
64. Alam, M.M., Malik, H., Khan, M.I., Pardy, T., Kuusik, A., Moullec, Y.L.: A survey on the roles of communication technologies in IoT-based personalized healthcare applications. IEEE Access **6**, 36611–36631 (2018)
65. Luo, E., Bhuiyan, M.Z., Wang, G., Rahman, M.A., Wu, J., Atiquzzaman, M.: PrivacyProtector: privacy-protected patient data collection in IoT-based healthcare systems. IEEE Commun. Mag. **56**, 163–168 (2018)

A Comparative Analysis of Image Denoising Problem: Noise Models, Denoising Filters and Applications

Subrato Bharati, Tanvir Zaman Khan, Prajoy Podder, and Nguyen Quoc Hung

Abstract Noise reduction is a perplexing undertaking for the researchers in digital image processing and has a wide range of applications in automation, IoT (Internet of Things), medicine, etc. Noise generates maximum critical disturbances as well as touches the medical images quality, ultrasound images in the field of biomedical imaging. The image is normally considered as a gathering of data and the existence of noises degradation the image quality. It ought to be vital to reestablish the original image noises for accomplishing maximum data from images. Digital images are debased through noise through its transmission and procurement. Noisy image reduces the image contrast, edges, textures, object details, and resolution, thereby decreasing the performance of postprocessing algorithms. This paper mainly focuses on Gaussian noise, salt and pepper noise, uniform noise, speckle noise. Different filtering techniques can be adapted for noise declining to improve the visual quality as well as a reorganization of images. Here four types of noises have been undertaken and applied to process images. Besides linear and nonlinear filtering methods like Gaussian filter, median filter, mean filter and Weiner filter applied for noise reduction as well as estimate the performance of filter through the parameters like mean square error (MSE), peak signal to noise ratio (PSNR), average difference value (AD)

S. Bharati · P. Podder (✉)
Institute of Information and Communication Technology, Bangladesh University of Engineering and Technology, Dhaka 1205, Bangladesh
e-mail: prajoypodder@gmail.com

S. Bharati
e-mail: subratobharati1@gmail.com

T. Z. Khan
Department of Information and Communication Engineering (ICE), Noakhali Science and Technology University (NSTU), Noakhali 3814, Bangladesh
e-mail: tzkhan19@gmail.com

N. Q. Hung
Department of Information Technology, School of Business Information Technology, University of Economics Ho Chi Minh City, Ho Chi Minh City 70000, Vietnam
e-mail: hungngq@ueh.edu.vn

A. E. Hassanien et al. (eds.), *Cognitive Internet of Medical Things for Smart Healthcare*, Studies in Systems, Decision and Control 311,
https://doi.org/10.1007/978-3-030-55833-8_3

and maximum difference value (MD) to diminish the noises without corrupting the medical image data.

Keywords Gaussian noise · Speckle noise · Salt and pepper noise · Uniform noise · Noise filters · IoT · Medical imaging

1 Introduction

Noise illustrates undesirable data, which breaks down the picture quality. It is characterized as a procedure which influences the obtained picture. Noise is generally introduced into pictures while exchanging and gaining them. Image noise causes an irregular variety of contrasts in the images. Images are influenced by different sorts of motion, for example, Gaussian noise generates via regular sources including particles thermal vibration as well as separate nature of radiation of warm object [1]. Exponential noise, speckle noise are a complex phenomenon, which corrupts picture value with the presence of backscattered wave which begins from numerous microscopic scattered reflections that going over interior organs besides create it more troublesome for the spectator to distinguish fine element of the image in investigative checkups [2]. Salt-and-pepper noise [3] is similarly mentioned as information drop noise. The picture is not completely disrupted by salt and pepper noise rather than some pixel qualities that are altered through the noise. Despite a noisy image, there is a plausibility of some neighbor's un-change. In the case of data transmission, this noise is found. Image pixel qualities are removed by defiled values of pixel either greatest "or" least pixel value specifically, 0 "otherwise" 255 individually. If transmission bits are 8. Poisson noise [4] and periodic noise are produced from gadgets obstructions, particularly in power signal during image procurement. This noise has uncommon qualities corresponding spatially reliant besides sinusoidal in an environment at products of a particular frequency. Periodic noise can be advantageously removed by utilizing a notch filter or narrowband rejects filter [5]. The presence of uniform noise is essential in the amplitude quantization procedure. It introduces due to the coversion or transformation of analog data into digital data. In this noise model, the signal to noise ratio (SNR) is restricted by the least and highest pixel esteem [6]. The principle sort of noise that is happening during the picture procurement is called Gaussian noise. Then again salt and pepper noise is, for the most part, presented while transmitting image information over an insecure correspondence filter. Along these lines, it is required to dispose of the noise from the picture to get exact information and degenerate the visual nature of the pictures. There are different kinds of filters use for picture noise decrease. For example, geometric mean filter, harmonic mean filter, median filter, Wiener filter, midpoint filter, max and min filter, alpha-trimmed mean filter, adaptive filter, bandpass filter, notch filter, band reject filter and so on. With a specific end goal to expel noise and enhance visual quality in this structure four dissimilar filters are utilized. Besides filter output data are evaluated utilizing five filters of superiority measures parameters. Histograms of these

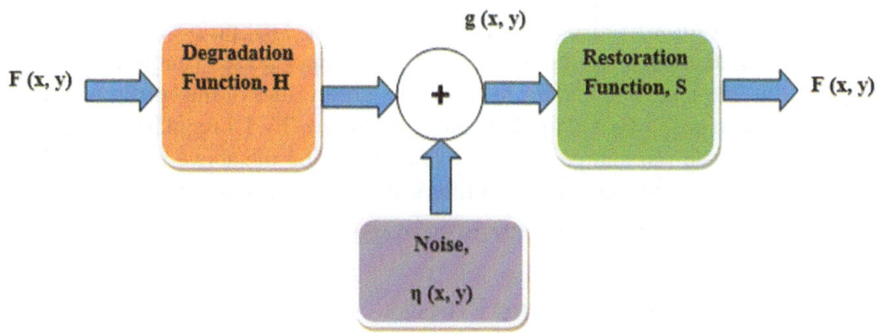

Fig. 1 Image degradation and restoration model

loud pictures are additionally discovered. The state of the histogram of the images have been demonstrated the kind of noise present in the images. Figure 1 demonstrates the image degradation as well as the restoration process where the input image is f(x,y) and g(x,y)is the noisy image. The image obtained after the restoration block is called the filtered image [7].

The paper is oriented in this fashion; related works are described in Sect. 2. The outline of the working process is described in Sect. 3. Types of image noise are described in Sect. 4, under this division Gaussian, speckle, salt and pepper and uniform noises are illustrated. Section 5 designates the de noising filters like mean, Gaussian, median, and Wiener filters. Besides experimental results and conclusion are represented in Sects. 6 and 7 respectively.

2 Related Works

Image denoising problem plays an important role in many fields of life such as medicine, automation, control, agriculture, etc. In the era of Industry 4.0, with the development of IoT (Internet of Things), image denoising is an important stage to improve image quality to increase the performance of postprocessing algorithms based on images for systems of IoT including applications in medicine.

Several researchers already worked on reducing the image noise. In the field of medical imaging systems, de-noising is an important image processing tasks. If noise can be removed automatically from a noisy medical image then it will improve the quality of diagnosis. Therefore, a doctor gives necessary treatment observing that image. Thanh et al. [8] provided a review of the Poisson removal methods such as the modified ROF model, bilateral filter PURE-LET method, etc. in his paper as CT and X-ray image de-noising methods. Computed tomography (CT) and X-Ray imaging systems use the X radiation for capturing the images. So, they are usually corrupted by Poisson noise. Erkan et al. [9] proposed an Iterative Mean (IM) Filter for eliminating the salt-and-pepper noise. The mean gray values of noise-free pixels are used in IMF in a fixed-size window. Thanh et al. [10] proposed an Adaptive Frequency Median Filter to remove the salt and pepper noise. The proposed filter employs the frequency median for restoring the grey values of the corrupted pixels. Thanh et al. also proposed and implemented an adaptive TV-L1 regularization model for eliminating the salt and pepper noise [11]. Inspiring the above research works, we have implemented several types of filters to remove various types of noise.

3 Overview of the Working Process

Figure 2 summarizes the working procedures of Gaussian, median, mean and Wiener filter. These are used in this working diagram for removing the undesired noise from the image.

4 Types of Image Noise

4.1 Gaussian Noise

Gaussian noise is generally called enhancer noise or random variation impulsive noise. Gaussian noise is created (a) electronic circuit noise, (b) sensor noise because of high temperature, (c) sensor noise due to poor brightening [12, 13]. It is a sort of measurable noise where the sufficiency of the noise takes after Gaussian dissemination [14]. The amplifier noise of the digital cameras is often approximately additive white Gaussian noise (AWGN). It can occur due to the thermally excited electrons in an electrical conductor [15]. This noise has Gaussian amplitude distribution. Mathematically this noise can be expressed as,

$$P(z) = \frac{1}{\sqrt{2\pi}\sigma} e^{\frac{-(z-\mu)^2}{2\sigma^2}} \tag{1}$$

where μ is the mean. Gaussian distribution can be also defined as normal distribution whose probability density function (PDF) must be equal to the statistical noise.

4.2 Speckle Noise

In [13, 14] speckle noise is one kind of granular noise and the picture quality has been degraded by this speckle noise. The images which are acquired from medical imaging techniques. That images are spoiled by the speckle noise. Generally, speckle noise

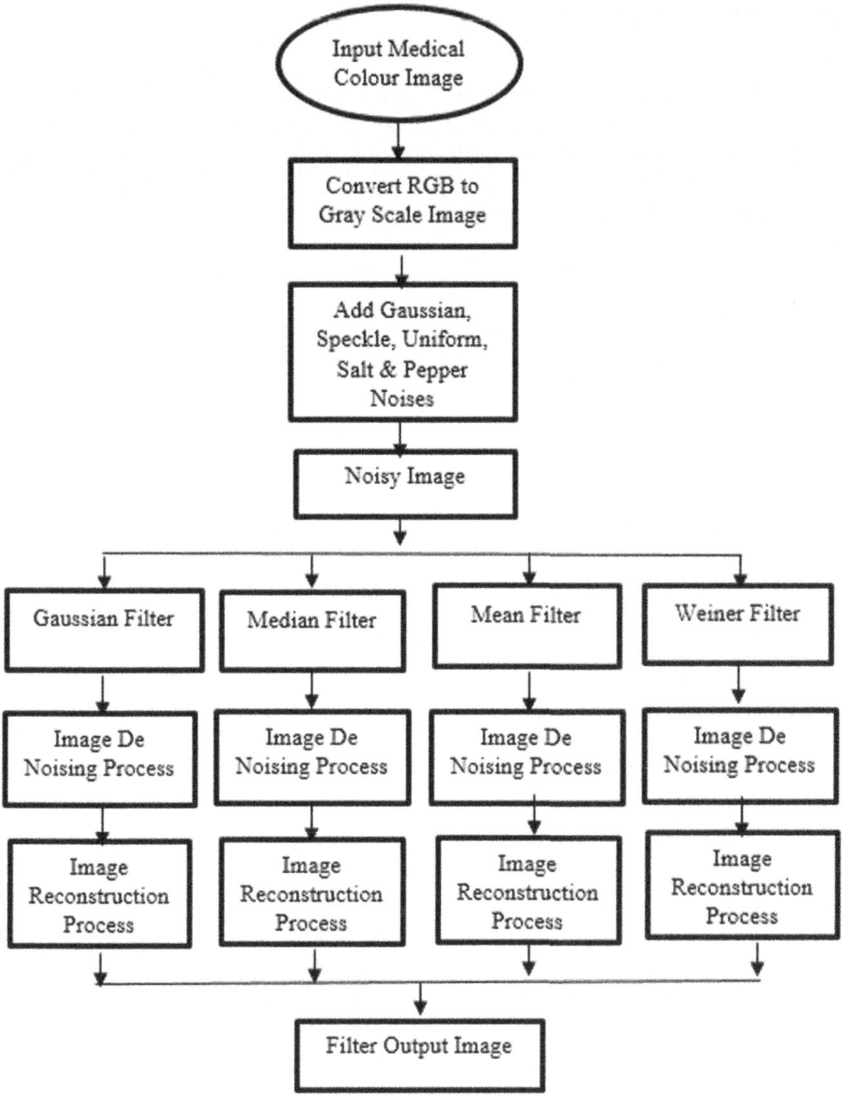

Fig. 2 Flowchart of working procedures

expands the mean gray near a native area and causing difficulties in medical images because of the coherent processing of backscattered signals [16].

4.3 Uniform Noise

Uniform or quantization noise is created by quantizing the pixels of a detected picture to various discrete levels. A signal is said to be an "Uniform" signal means that the signal comprises of the random values from a uniform (rectangular type) distribution. Therefore, every value of that signal in this range is equally likely. [12].

The PDF of uniform noise is given by,

$$P(z) = \begin{cases} \frac{1}{d-c}, & if\ c \le z \le d \\ 0, & otherwise \end{cases} \tag{2}$$

Mean value is specified by,

$$\mu = \frac{c+d}{2} \tag{3}$$

Variance is given by,

$$\sigma^2 = \frac{(d-c)^2}{12} \tag{4}$$

4.4 Salt and Pepper Noise

Salt and pepper noise is called impulse noise. It can generally be named as a spike noise [17]. The salt and pepper (SnP) noise is brought about by sudden and sharp disorder in the image signal. It is apparent as arbitrarily scattered dark or white (or together) pixels above the image. The SnP noise affected pixel has two possible values. The value of white pixel or salt pixel is 255. The value of black or pepper pixel is 0. The noisy (SnP) images must be a 8-bit grayscale image. It is digitized as great quality in the image. Impulse noise contained image has dark pixels in the bright region [12]. The PDF of (bipolar) impulse noise can be communicated as [12].

$$p(a) = \begin{Bmatrix} P_c, & for\ \ a = c; \\ P_d, & for\ \ a = d; \\ 0, & otherwise \end{Bmatrix} \tag{5}$$

If $d > c$, gray level b will perform as a light dot in the image. On the contrary, level c will act similar to a dark dot. c is 0 (black) and c is 255 (white) for the 8 bit image. The impulse noise is called uni-polar when P_c or P_d is zero.

5 Image Denoising Filters

In last 20 years, the researchers are continuously trying to improve efficient image denoising algorithms These algorithms aim at restoring a reasonable amount of information from the distorted and corrupted image at the time of preserving the fine features and edges. As technical complexities in IoT driven image acquisition devices are rising, digital sensors have a tendency to increase the image pixel resolution. But these sensors with its aperture size become more prone to noise. As a result, it creates the necessity of implementing the image denoising algorithms. Software-oriented approaches are mostly device independent. They are broadly employed compared to the hardware and optical systems [18, 19]. For example, noise free images acan be considered a prerequisite condition in various areas such as investigation of medical images, detection of lung diseases from X-ray images with the help of deep learning [20], COVID-19 detection from CT or X-ray images [21], real time object detection from images [22], analysis of signals (video, sound, voice), information removal, radio space science, etc. Every application has its exceptional necessities. For instance, noise clearance in medical data requires particular consideration since information taken from the medical signal (ECG, EEG [23], and so on.) is exceptionally touchy. Noise-free images are required for line, point, and edge discovery. In image pre-processing section, smoothing and enhancing or sharpening the desired image can be considered two basic filtering operation.

High-frequency parts can be contained in the images. Smoothing filters can assume an imperative part to suppress those high frequencies [24]. Then again, sharping filters are utilized to smooth image low frequency, i.e. improving or recognizing picture edges. Image rebuilding and upgrade methodologies can be characterized into spatial domain and frequency domain classifications. This arrangement is typically in light of changing the Fourier transformation of an image. Noise removal is perplexing in the frequency domain when contrasted with the spatial domain. The spatial domain noise expulsion involves fewer meting out time. Noise removal algorithms ought to give a tasteful measure of noise evacuation furthermore safeguard the edges. For fulfilling expressed conditions there are two sorts of filters: linear and non-linear with their noteworthy favorable circumstances and disservices. The linear filters have the benefit of speedier planning and the shortcoming of not sparing edges. A fact in non-linear filters has the upside of securing edges and the downside of slower preparation [25]. In this work, four kinds of filters have been talked about underneath which will be utilized further to obliteratethe noise information from the main image.

5.1 Median Filter

Order-statistics filter also known as median filter, which exchanges the estimation of a pixel by the middle of the gray levels in the region of that pixel. The median is a rank command statistic and in intelligence, the mainstream of the included pixel values determine the result [26]. The expression of the median filter,

$$f'(x, y) = \underset{(s,t)\in S_{xy}}{median}\{g(s, t)\} \tag{6}$$

The first estimation of the pixel is incorporated into the calculation of the median. Median filters are extremely standard for assured sorts of arbitrary noise. They give astonishing noise diminish capacities than a linear smoothing filter of comparative size [27].

The median is figured by first sorting all the pixel values from the window in numerical order. A while later supplanting the pixel being considered with the inside (middle) pixel value. Median filters are stable for the bipolar and unipolar impulse noise. Median filters are mainly reasonable within the sight of both unipolar and bipolar impulse noise. Trimmed median noise removal filter can play an effective role for images corrupted by scratches caused by the human, environment [28], etc.

5.2 Gaussian Filter

Gaussian filter is a linear smoothing filter, where the weights are chosen for the smoothing purpose according to the outline of the function of Gaussian. Gaussian filter in the nonstop space can be defined by the following equation,

$$h(m, n) = \left(\frac{1}{\sqrt{2\pi\sigma}}e^{-\frac{m^2}{2\sigma^2}}\right) \times \left(\frac{1}{\sqrt{2\pi\sigma}}e^{-\frac{n^2}{2\sigma^2}}\right) \tag{7}$$

One-dimensional Gaussian filter has an impulse response,

$$g(x) = \sqrt{\frac{a}{\pi}}e^{-ax^2} \tag{8}$$

This equation can also be uttered with the standard deviation as a parameter,

$$g(x) = \frac{1}{\sqrt{2\pi}\cdot\sigma}e^{-\frac{x^2}{2\sigma^2}} \tag{9}$$

where, the standard deviations are mentioned in their physical units, e.g. time and frequency in seconds and Hertz [29, 30].

5.3 Mean Filter

Mean filtering is the simplest way to decrease the amount of intensity deviation between one and next pixel for smoothing image [14]. The arithmetic mean-filter can be expressed by [31].

$$f'(x, y) = \frac{1}{mn} \sum_{(s,t)\in S_{xy}} g(s, t) \tag{10}$$

At this time, g is the disrupted image, r and c are the rows and columns coordinates correspondingly within a window size of m × n besides the filtered image is $\hat{f}(x, y)$. However, the geometric mean filter is a variation of the arithmetic mean filter, which calculated appearance can be known as follows,

$$f'(x, y) = \left[\prod_{(r,c)\in W} g(r, c) \right]^{1/mn} \tag{11}$$

5.4 Wiener Filter

The Wiener filter is a standout among the most fundamental methodologies for noise diminishment. Additive noise is removed by Wiener filtering. It also inverts the blurring of the image [32]. The Time-domain Wiener filter is derived by minimizing the MSE between the image of interest and its assessment [33, 34]. The Wiener filter is ideal in minimum mean square error sense for recuperating noise image. The detected image g(x,y) is thought to come about because of the entirety of the stationary noise n(x, y) and original image f(x, y).

$$g(x, y) = f(x, y) + n(x, y) \tag{12}$$

where noise is spectrally white with zero mean and variance σ^2. The transfer function of the Wiener filter is,

$$H(u, v) = \frac{P_f(u, v)}{P_f(u, v) + \sigma^2} \tag{13}$$

For image retrieval noise adaptive Wiener filtering is expressed as [27],

$$f'(x, y) = m_f(x, y) + \frac{\sigma_f^2(x, y)}{\sigma_f^2(x, y) + \sigma_n^2(x, y)} \big(g(x, y) - m_f(x, y)\big) \tag{14}$$

6 Experimental Results

Matrix laboratory software (MATLAB) has been used to analyze the performances of noise removal filters on four different types of noise.

Figures 3, 4, 5, 6 and 7 refer to the effect of filtering consecutively for removing the Gaussian noise, speckle noise, uniform noise, salt and pepper noise from medical images where, Figs. 3a, 4a, 5a, 6a, 7a signifies the original images.

Figures 3b, 4b, 5b, 6b represents sequentially Gaussian noise, speckle noise, uniform noise, salt and pepper noise affected images.

Figures 3c, 4c, 5c, 6c, 7c represents a median filter for removing the Gaussian noise, speckle noise, uniform noise, salt and pepper noise and speckle noise from medical images.

Figures 3d–f, 4d–f, 5d–f, 6d–f, 7d–f represents consecutively the performance of Gaussian, mean, Weiner filter for removing the Gaussian noise, speckle noise, unifrom noise, salt and pepper noise and speckle noise from medical images.

Figures 8, 9, 10 and 11 evaluate the analysis of the performance of filters such as Gaussian, median, mean, and Weiner through MSE, PSNR, AD as well as MD.

6.1 *Mean Square Error (MSE)*

Mean square error can be calculated by the following equation,

$$MSE = \frac{1}{MN}\sum_{x=1}^{M}\sum_{y=1}^{N}[g(x, y) - f(x, y)]^2 \tag{8}$$

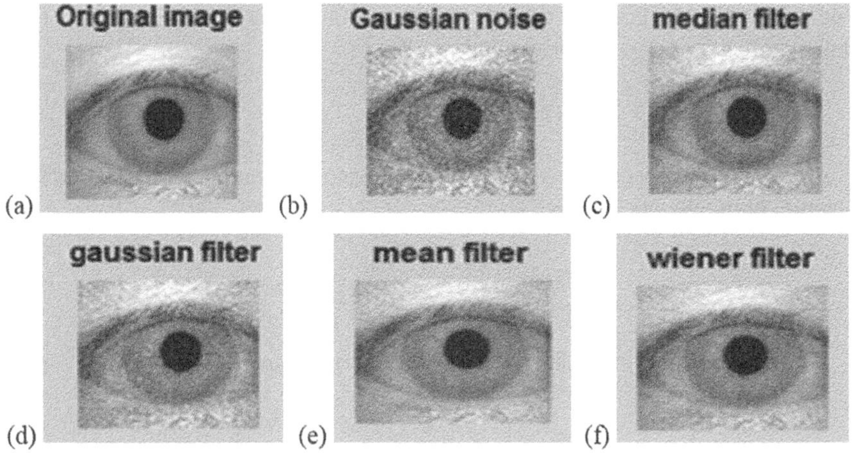

Fig. 3 Result of filtering for removing Gaussian noise

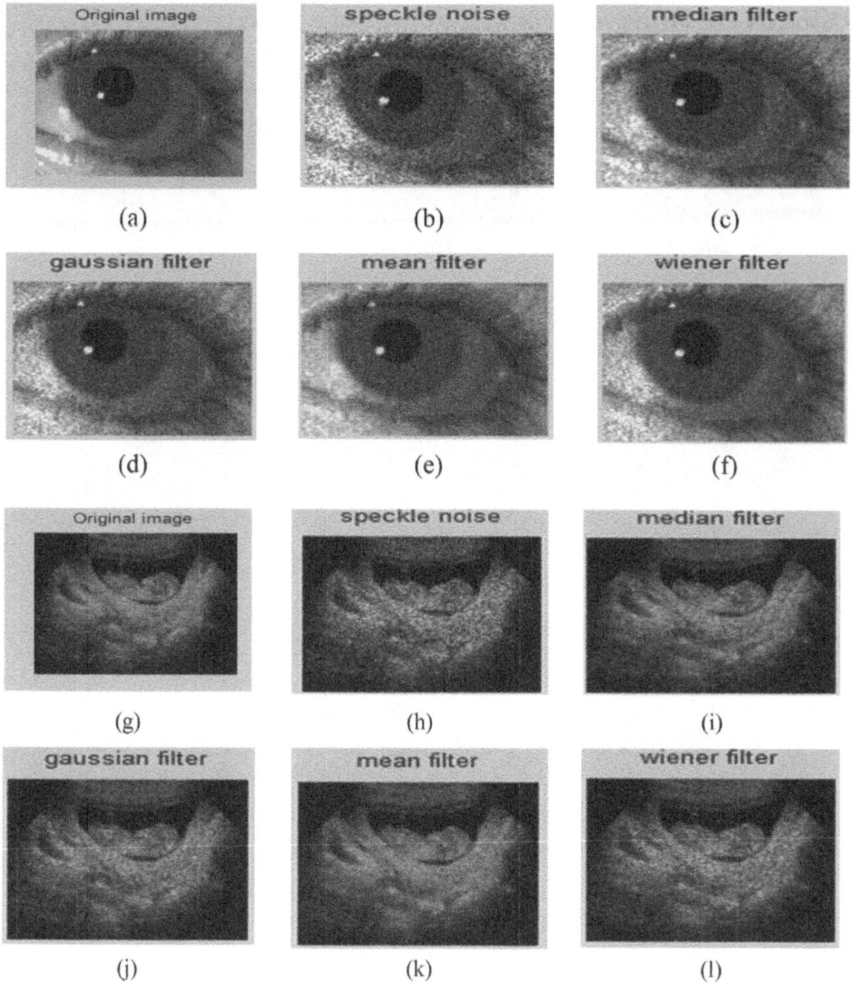

Fig. 4 Effect of filtering for removing speckle noise

where, M × N is the total number of pixels, g(x,y) is corrupted image, and f(x,y) is filtered image. The lowest value of mean square error represents the best-filtered image.

Table 1 and Fig. 8 show that the Gaussian filter is ideal for uniform noise than other noises reduction. Because of lower MSE, AD, MD as well as higher PSNR value. Besides Gaussian noise reduction can be possible more by Gaussian filter than median, mean, and Weiner filters.

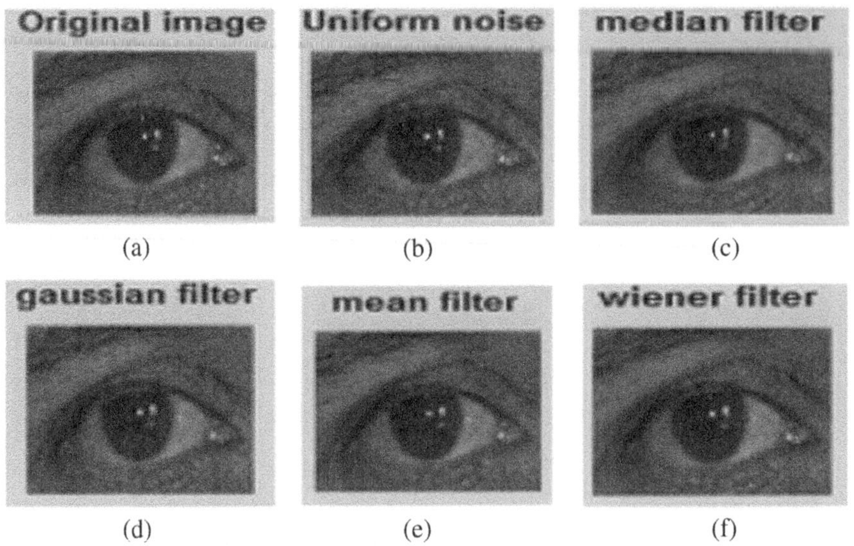

Fig. 5 Result of filtering for removing uniform noise

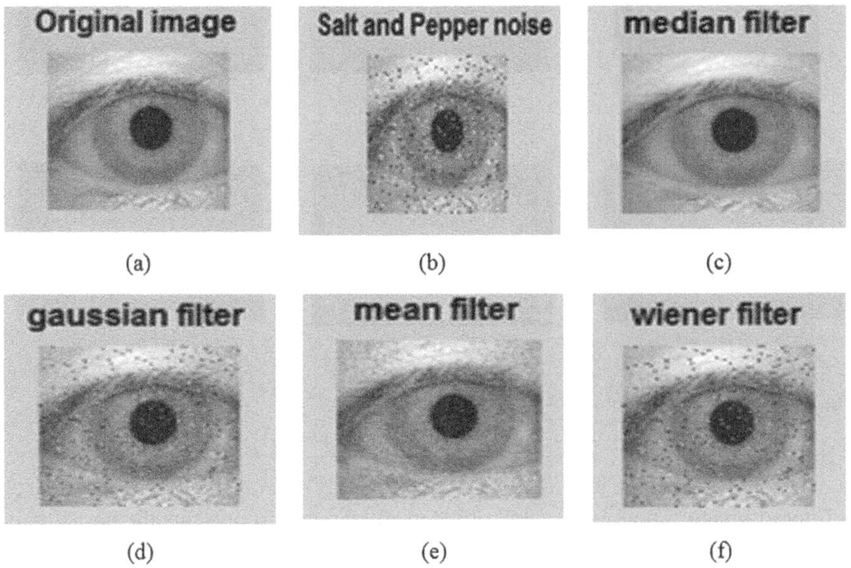

Fig. 6 Result of filtering for eliminating salt and pepper noise

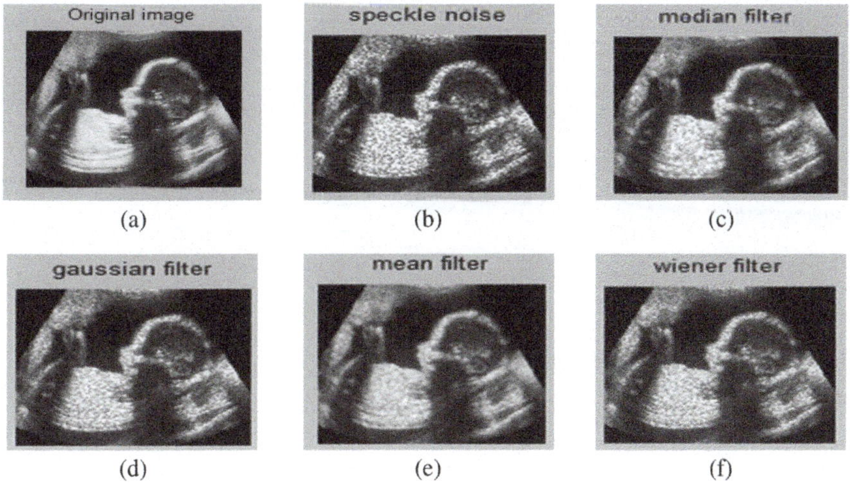

Fig. 7 Result of filtering for eliminating speckle noise from ultrasound image

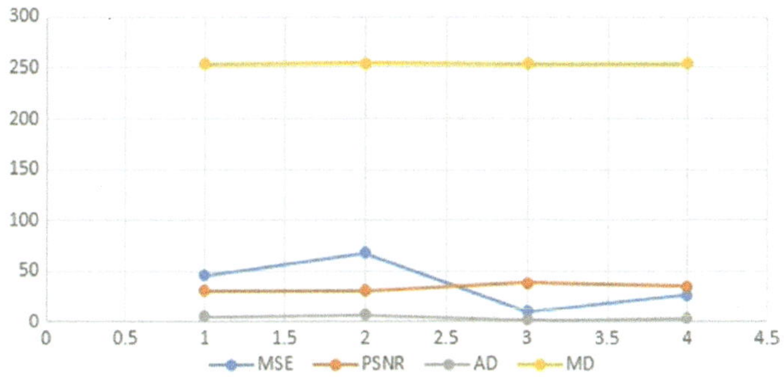

Fig. 8 Performance analysis of Gaussian filter

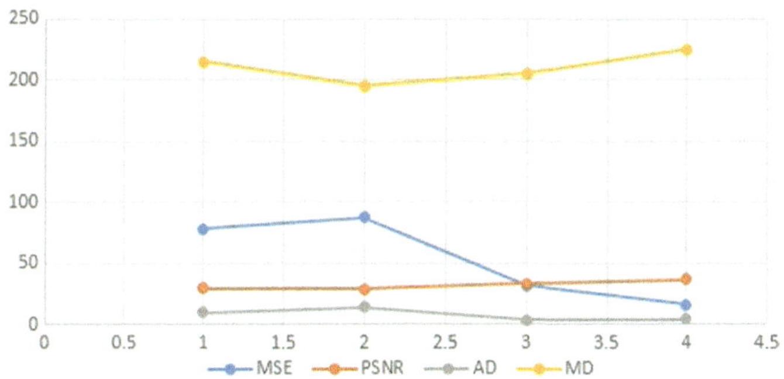

Fig. 9 Performance analysis of median filter

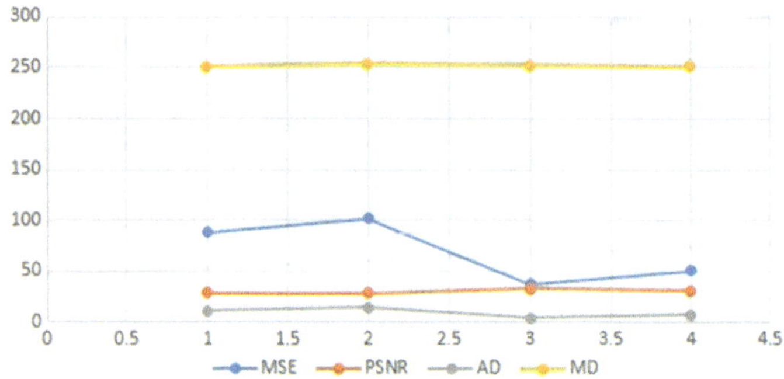

Fig. 10 Performance analysis of mean filter

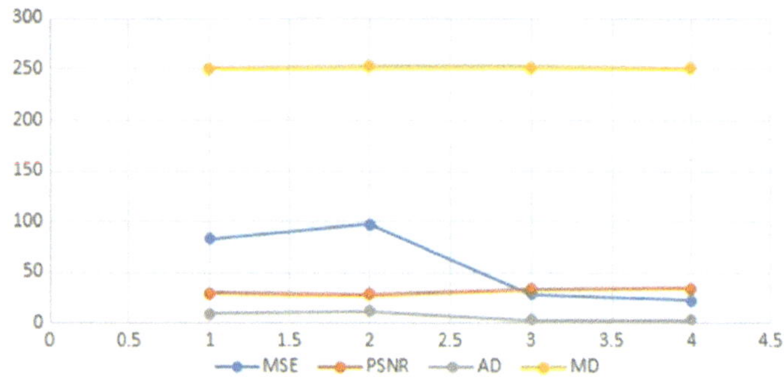

Fig. 11 Performance analysis of Weiner filter

Table 1 Gaussian filter performance analysis on different types of noise

Gaussian Filter	Types of noise			
	Gaussian	Speckle	Uniform	Salt and pepper
MSE	45.46	67.82	9.74	25.77
PSNR	30.39	29.92	37.89	34.04
AD	4.40	6.10	1.61	2.89
MD	254	255	254	254

6.2 *Peak Signal to Noise Ratio (PSNR)*

The Peak Signal to Noise Ratio (PSNR) can be considered by the following equation,

$$PSNR = 10\log_{10}\left[\frac{255 \times 255}{MSE}\right] \tag{9}$$

Table 2 Median filter performance analysis on different types of noise

Median Filter	Types of noise			
	Gaussian	Speckle	Uniform	Salt and pepper
MSE	78.14	87.29	31.37	15.80
PSNR	29.25	28.77	33.20	36.18
MAD	9.36	13.6	3.15	3.85
MD	215	195	205	225

Table 3 Mean filter performance analysis on different types of noise

Mean filter	Types of noise			
	Gaussian	Speckle	Uniform	Salt and pepper
MSE	87.81	101.12	36.56	50.91
PSNR	28.74	28.12	32.67	31.10
AD	10.37	14.22	4.10	6.97
MD	251	255	253	252

For image quality measurement, if the value of PSNR is very high for an image of a particular noise type then it is defined as the best quality image.

In the case of the median filter, salt and pepper noise is appropriate for best noise reduction performances for medical images [35, 36] due to the highest PSNR and lowest MSE value (from Table 2). However, the median filter works well for uniform noise reduction.

6.3 *Average Difference Value (AD)*

A lower value of average difference represents a cleaner image. The average difference can be calculated by the following equation,

$$AD = \frac{1}{MN}\sum_{x=1}^{M}\sum_{y=1}^{N}[g(x, y) - f(x, y)] \tag{10}$$

Table 3 shows speckle noise produces higher MSE values noise reduction whereas uniform, salt and pepper noise provides a smaller value of MSE to mean filter.

6.4 *Maximum Difference (MD) Value*

Maximum difference can be expressed by the following formula,

Table 4 Weiner filter performance analysis on different sorts of noise

Weiner filter	Types of noise			
	Gaussian	Speckle	Uniform	Salt and pepper
MSE	82.70	97.18	28.64	22.31
PSNR	29.14	28.30	33.59	34.67
AD	8.62	11.67	3.07	3.44
MD	251	254	253	252

$$MD = Max(|f(x, y) - g(x, y)|) \tag{11}$$

The quality of the picture is poor if the value of the maximum difference is large. Figures 8, 9, 10 and 11 illustrate the performances of Gaussian, median, mean and Weiner filter. Table 4 describes the performance of Weiner filter on the noisy image.

7 Conclusion

Noises are random in this environment as well as deterministic in the biomedical imaging scheme. It is difficult to remove noise present in the medical and ultrasound images, since the information about the variance of the noise may not be able to identify. This paper generally motivated on the various noise and noise suppression filters, i.e., Gaussian, median, mean, Weiner. The working process establishes main parameters that have been pragmatic on medical images (for example Iris and Ultrasound image). The investigational opinion under MSE, PSNR, AD, and MD show efficient filter performance at various noises. A relative study has been motivated among those filters to successfully de-noising the images. The conclusion of this study exhibits that uniform noise provides relatively better performance for reduction among the other noises in terms of PSNR, MSE, and visual quality. Besides Gaussian filter's noise removal performances are comparatively better than the median, mean, and Weiner filters in case of uniform, Gaussian, speckle, salt and pepper noises. CT or X-ray images must be noise free at the time of detecting COVID-19 from the input images with the help of various deep learning algorithms.

References

1. Boyat, A., Joshi, B.K.: Image denoising using wavelet transform and median filtering. In: IEEE Nirma University International Conference on Engineering, Ahmedabad (2013)
2. Kumbhakarna, V., Patil, V.R., Kawathekar, S.: Review on speckle noise reduction techniques for medical ultrasound image processing. Int. J. Comput. Tech. **2**(1) (2015)
3. Joshi, A., Boyat, A.K., Joshi, B.K.: Impact of wavelet transform and median filtering on removal of salt and pepper noise in digital images. In: International Conference on Issues and Challenges

in Intelligent Computing Techniques (ICICT) (2014). https://doi.org/10.1109/ICICICT.2014.6781389
4. Rodrigues, I., Sanches, J.: Denoising of medical images corrupted by Poisson noise. In: 15th IEEE International Conference on Image Processing 2008. https://doi.org/10.1109/ICIP.2008.4712115
5. Hamd, M., Rasool, R.: Dynamic restoration of periodic noisy image using: upper-half spectrum. In: International Conference on Industrial Informatics and Computer Systems (CIICS) (2016). https://doi.org/10.1109/ICCSII.2016.7462407
6. Panda, C.S., Patnaik, S.: Filtering and performance evaluation for restoration of grayscale image corrupted by salt and pepper noise using low pass filtering schemes. In: 2009 Second International Conference on Emerging Trends in Engineering and Technology, ICETET, Nagpur, pp. 940–945 (2009)
7. Boncelet, C.: Image noise models In: Bovik, A.C. (ed.) Handbook of Image and Video Processing. Academic Press. (2005). ISBN 0-12-119792-1.
8. Thanh, D., Surya, P., Hieu, L.M.: A review on CT and X-ray images denoising methods. Informatica **43**(2), 151–159 (2019)
9. Erkan, U., Thanh, D.N.H., Hieu, L.M., Enginoglu, S.: An iterative mean filter for image denoising. IEEE Access **7**, 167847–167859 (2019)
10. Erkan, U., Enginoglu, S., Thanh, D.N.H., Hieu, L.M.: Adaptive frequency median filter for the salt-and-pepper denoising problem. IET Image Proc. (2019). https://doi.org/10.1049/iet-ipr.2019.0398
11. Thanh, D.N.H., Thanh, L.T., Hien, N.N., Prasath, S.: Adaptive total variation L1 regularization for salt and pepper image denoising. Optik 163677 (2019). ISSN 0030-4026. https://doi.org/10.1016/j.ijleo.2019.163677
12. Thanh, D.N., Hai, N.H., Prasath, V.S., Hieu, L.M., & Tavares, J.M.R.:A two-stage filter for high density salt and pepper denoising. Multimedia tools and applications. https://doi.org/10.1007/s11042-020-08887-6 (2020)
13. Boyat, A.K., Joshi, A.: Noise models in digital image. Signal Image Process. Int. J. (SIPIJ) **6**(2) (2015)
14. Chan, R.H., Ho, C.-W., Nikolova, M.: Salt-and-pepper noise removal by median-type noise detectors and detail-preserving regularization. IEEE Trans. Image Process. **14**(10) (2005)
15. Goyal, B., Dogra, A., Agrawal, S., Sohi, B.S.: Noise issues prevailing in various types of medical images. Biomed. Pharmacol. J. **11**(3) (2018)
16. Podder, P., Hasan, M.M.: A meta study of reduction of speckle noise adopting different filtering techniques. In: 2016 3rd International Conference on Electrical Engineering and Information Communication Technology (ICEEICT), Dhaka, pp. 1–6 (2016)
17. Thanh, D.N.H., Dvoenko, S.D.: A method of total variation to remove the mixed Poisson-Gaussian noise. Pattern Recogn. Image Anal. **26**(2), 285–293 (2016)
18. Huang, Z., Zhang, Y., Li, Q., Zhang, T., Sang, N., Hong, H.: Progressive dual-domain filter for enhancing and denoising optical remote-sensing images. IEEE Geosci. Remote Sens. Lett. **15**(5), 759–763 (2018)
19. Shao, L., Yan, R., Li, X., Liu, Y.: From heuristic optimization to dictionary learning: a review and comprehensive comparison of image denoising algorithms. IEEE Trans. Cybern. **44**(7), 1001–1013 (2014)
20. Bharati, S., Prajoy Podder, M., Mondal, R.H.: Hybrid deep learning for detecting lung diseases from X-ray images. Inf. Med. Unlocked **20**, 100391 (2020). https://doi.org/10.1016/j.imu.2020.100391
21. Rubaiyat Hossain Mondal, M., Bharati, S., Podder, P., Podder, P.: Data analytics for novel coronavirus disease. Inf. Med. Unlocked **20**, 100374 (2020). https://doi.org/10.1016/j.imu.2020.100374
22. Hossain, F. Raihan-Al-Masud, M., Rubaiyat Hossain Mondal, M.: Effectiveness of deep learning in real time object detection. Int. J. Comput. Appl. **176**(41), 55–60 (2020)
23. Sadiq Iqbal, M., Nasim Akhtar, M., Shahariar Parvez, A.H.M., Bharati, S., Podder, P.: Ensemble learning-based EEG feature vector analysis for brain computer interface. In: Suma, V., Bouhmala, N., Wang, H. (eds.) Evolutionary Computing and Mobile Sustainable Networks. Lecture

Notes on Data Engineering and Communications Technologies, vol. 53. Springer, Singapore (2021). https://doi.org/10.1007/978-981-15-5258-8_88
24. Goyal, B., Dogra, A., Agrawal, S., Sohi, B.S., Sharma, A.: Image denoising review: from classical to state-of-the-art approaches. Inf. Fus. **55**, 220–244 (2020). ISSN 1566-2535. https://doi.org/10.1016/j.inffus.2019.09.003
25. Gupta, G.: Algorithm for image processing using improved median filter and comparison of mean and improved median filter. IJSCE **1**(5) (2011). ISSN: 2231–2307, November 2011
26. Chakre, R., Thapa, P., Siddharth, R., Baruah, U.: An experimental study on image de-noising filters. Int. J. Emerg. Technol. Adv. Eng. **3**(2) (2013)
27. Thanh, D.N., Hien, N.N., Kalavathi, P., Prasath, V.S.: Adaptive switching weight mean filter for salt and pepper image denoising. Proc. Comput. Sci. **171**, 292–301 (2020).
28. Thanh, D.N.H., Son, N.V., Prasath, V.B.S.: Distorted image reconstruction method with trimmed median. In: 2019 3rd International Conference on Recent Advances in Signal Processing, Telecommunications & Computing (SigTelCom), Hanoi (2019)
29. Thanh, D.N., Prasath, V.S., Dvoenko, S.: An adaptive method for image restoration based on high-order total variation and inverse gradient. Sig. Image Video Proc. 1–9 (2020). https://doi.org/10.1007/s11760-020-01657-9
30. Bharati, S., Rahman, M.A., Mandal, S. and Podder, P.: Analysis of DWT, DCT, BFO & PBFO Algorithm for the Purpose of Medical Image Watermarking. In 2018 International Conference on Innovation in Engineering and Technology (ICIET), IEEE, pp. 1-6 (December, 2018). https://doi.org/10.1109/CIET.2018.8660796
31. Shi Zhong, G.: Image de-noising using wavelet thresholding and model selection. In: International Conference on Image Processing, vol 3, Sept 2000
32. Puri, D., Mishra, S.K.: Analysis of image restoration techniques at different noises. Int. J. Comput. Sci. Inf. Technol. **8**(3), 439–442 (2017)
33. Kabir, M.A., Mondal, M.R.H.: Edge-based transformation and entropy coding for lossless image compression. In: 2017 International Conference on Electrical, Computer and Communication Engineering (ECCE), pp. 717–722. IEEE (February 2017). https://doi.org/10.1109/ECACE.2017.7912997
34. Chen, H., Cen, Z., Wang, C., Lan, S., Li, X.: Image restoration via improved Wiener filter applied to optical sparse aperture systems. Optik **147**, 350–359 (2017). ISSN 0030-4026. https://doi.org/10.1016/j.ijleo.2017.08.102
35. Khamparia, A., Saini, G., Pandey, B., et al.: KDSAE: Chronic kidney disease classification with multimedia data learning using deep stacked autoencoder network. Multimed. Tools Appl. (2019). https://doi.org/10.1007/s11042-019-07839-z
36. Chouhan, V., Singh, S.K., Khamparia, A., Gupta, D., Tiwari, P., Moreira, C., Damaševičius, R., de Albuquerque, V.H.C.: A novel transfer learning based approach for pneumonia detection in chest X-ray Images. Appl. Sci. **10**, 559 (2020)

Applications and Challenges of Cloud Integrated IoMT

Subrato Bharati, Prajoy Podder, M. Rubaiyat Hossain Mondal, and Pinto Kumar Paul

Abstract The Internet of Medical Things (IoMT) refers to the medical devices and applications that connect healthcare information technology (IT) systems via computer networks. This chapter focuses on different aspects including the strengths, weakness, prospects and challenges of the IoMT integrated cloud computing. First of all, a gap analysis has been performed which indicates that there are some limitations in the existing computation capability, communication protocols, scalability, infrastructure, data security, etc. of IoMT. Secondly, different characteristics of cloud computing including resource polling, on demand services, access network, and security as well as privacy issues are discussed. Thirdly, a framework for IoT healthcare network (IoThNet) is presented which illustrates how hospitals at access layer can collect user information at data persistent layer. Next, the local storage and cloud storage platforms of IoThNet are briefly explained. A communication system is described then where a patient is monitored by the transmission of medical data via the wearable sensors on the patient. We propose a cloud integrated IoMT framework and compare it with the existing frameworks reported in the literature. Patients and their relatives, doctors can use this framework to get the health status of the patients and get alert in case of emergency conditions. Next discussion is provided on a number of healthcare services for instance adverse drug reaction, and on healthcare applications such as glucose level sensing and wheelchair management. A description is also provided on how IoMT can help support different diseases with the help

S. Bharati · P. Podder (✉) · M. R. H. Mondal
Institute of Information and Communication Technology, Bangladesh University of Engineering and Technology, Dhaka 1205, Bangladesh
e-mail: prajoypodder@gmail.com

S. Bharati
e-mail: subratobharati1@gmail.com

M. R. H. Mondal
e-mail: rubaiyat97@yahoo.com

P. K. Paul
Department of CSE, Ranada Prasad Shaha University, Narayanganj 1400, Bangladesh
e-mail: pinto.kumar07@gmail.com

A. E. Hassanien et al. (eds.), *Cognitive Internet of Medical Things for Smart Healthcare*, Studies in Systems, Decision and Control 311,
https://doi.org/10.1007/978-3-030-55833-8_4

of sensors for example, glucose, pulse, temperature, blood pressure, heart rate, force, etc. sensors. Finally, smartphone applications (apps) for diagnosis, drug reference, medical education and clinical communication are reported.

Keywords Medical · Cloud · Network · Healthcare · Sensors · IoT · IoMT · Security

1 Introduction

The Internet of Things (IoT) is an emerging technology that refers to the system of interconnected devices collecting and sharing real time data without human intervention. It can control the healthcare system, smart cities, security, waste management, emergency services, traffic congestion, industrial control, logistics, and retails. Nowadays, infrastructure of the Internet is extended with some benefits and interconnected with other services. Benefits usually include advanced connectivity, services and, systems that drives beyond machine-to-machine (M2M) developments [1]. As a result, introducing automation is feasible in practically all fields. The IoT offers a wide range of applications with its solutions [1–3].

Health care and medical care present one of the attractive application areas for the Internet of Medical Things (IoMT) [4]. IoMT is based on machine to machine communication where medical devices or machines are equipped with wireless connectively to transfer data. The IoMT has the potential for various medical applications i.e. fitness programs, remote health monitoring, elderly care, and chronic diseases. Home medication, and compliance with treatment are another significant and potential applications. As a result, various sensors, and medical devices, imaging devices and diagnostic can be observed as smart devices creating a primary part of the IoMT.

IoMT-based healthcare systems are estimated to increase the quality of life, reduce costs, and develop the user's knowledge. The IoMT has the potential to decrease device interruption through remote provision from the provider healthcare perspective. Moreover, the IoMT can properly detect optimum times for replacing supplies for several devices for their continuous and smooth operation. Furthermore, the IoMT offers for the effective scheduling of inadequate resources by confirming their best use as well as service of new patients.

Cloud computing provides computing services i.e. databases, servers, software, data analytics, and networking over the Internet to deliver flexible resources, faster deployment, and economies of scale [5, 6]. IoMT devices can connect with cloud services for the storage and processing of medical data.

According to the report published by CISCO [7], 500 billion devices will be approximately connected by 2030. IoT market study performed by Statista [8] in 2017 indicated that the IoT market will be valued as high as 8.9 trillion US dollar in the world by 2020 where 7% of the total market value comes from the healthcare sector. So, IoMT is playing a crucial role in healthcare sector.

A short descriptive summary is conducted in this chapter about the IoT integrated cloud computing for developing the healthcare system. Various categories of IoT and cloud computing in the field of digital healthcare are covered such as the integration of IoT and cloud computing structure, platforms providing healthcare applications in order to establish the communication between IoT and cloud computing backbone. Moreover, an IoMT framework is proposed and compared with the existing frameworks reported in the literature.

2 A Gap Analysis of Internet of Things in Healthcare Environment

The future condition of IoT integrated system and its contribution in healthcare can be evaluated by gap analysis. In this section, gap analysis is introduced in order to identify the strong points, weaknesses, prospects, and challenges of the fusion of IoT and cloud computing in the medical field. According to Borgia et al., heterogeneity and scalability are two important requirements of IoT [9]. So, some stages of deep analysis should be taken into consideration for facing that requirements. Those are:

(i) Architectural analysis
(ii) Communication system
(iii) Object name
(iv) Coding mapping services

Cost can also be considered an important factor in the analysis process. For minimizing the cost of each stage and the total integrated system, the development cost, installation cost, and maintenance cost should be prioritized and guaranteed. There are some limitations of IoT in providing solution of administrative and financial problems and it plays a small role in direct care for patients. However, the IoMT is playing an important role in delivering a noteworthy and proficient care for patients. Scalability is also an important issue for the heterogeneous devices operating in IoT platform. Objects are IP-enabled in the IoT interconnected system. So, services of IoT in this case can be implemented by integrating objects into the cloud environment. Then the whole system will become flexible. A gap analysis of IoT paradigm is presented in [10–13]. This analysis is based on the integration of sensing and actuating technologies, data sharing and processing. IoT driven technique can also be effective in deep learning for detection and classification of cancer [14, 15]. Figure 1 summarizes in more details the gap analysis of IoT platform.

3 Cloud Computing in Healthcare Systems

Cloud computing is a technology designed to provide an enormous number of resources as well as computing services via a networked platform including the

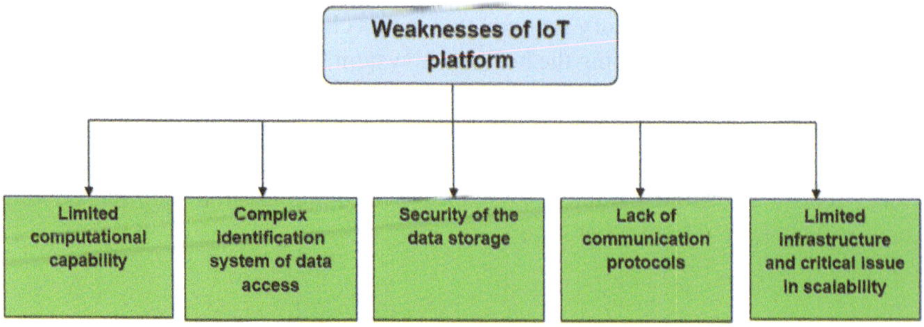

Fig. 1 A gap analysis of IoT platform

Web. In simple terms, cloud computing refers to the use of servers on the Internet for storing, managing, and processing data. It provides several benefits to the prospective users i.e. virtual hardware and collaboration software, virtual storage, and virtual servers. The term of cloud can be defined as a symbol for the Internet. Various researchers offered novel emerging information and communication technology (ICT) service model cloud computing [16]. According to the paper of [17], there are 22 descriptions of cloud computing. Cloud computing is distributed into three core services: Platform as a Service, Infrastructure as a Service (IaaS), and Software as a Service (SaaS) [18]. There are four kinds of clouds. The first one is public cloud for the common public to purchase and access from service suppliers i.e. Amazon. The second one is private cloud. The third one is hybrid cloud used for both public and private clouds. The last one is community cloud. It is based on the development and membership of research societies with comparable interests [19, 20]. Figure 2 reviews the key aspects of cloud paradigm. It is a complete model of main characteristics. Figure 2 shows on-demand services that refer to the easy entrance by health and medical cares to an extensive range of information from different sources i.e. claims, electronic medical records (EMRs), laboratory data, and medication. Moreover, drug regimens of the treatment for adhering or chronic asthma are automatically notified doctors to missed or conflicting prescriptions. We illustrate briefly the key advantages of cloud computing applications for healthcare. The authors of [21] discussed the following issues:

(a) Customized service: The patients' requests can be automatically pleased without any human operator.
(b) The access of network: The patients' applications can run over various platforms by using electrical devices i.e. laptops, tablets, smartphones, etc.
(c) Remote access: The properties of the cloud i.e. software and hardware are unknown to patients. The access and services are delivered to the patients slightly without knowing the locations of data to store or access.

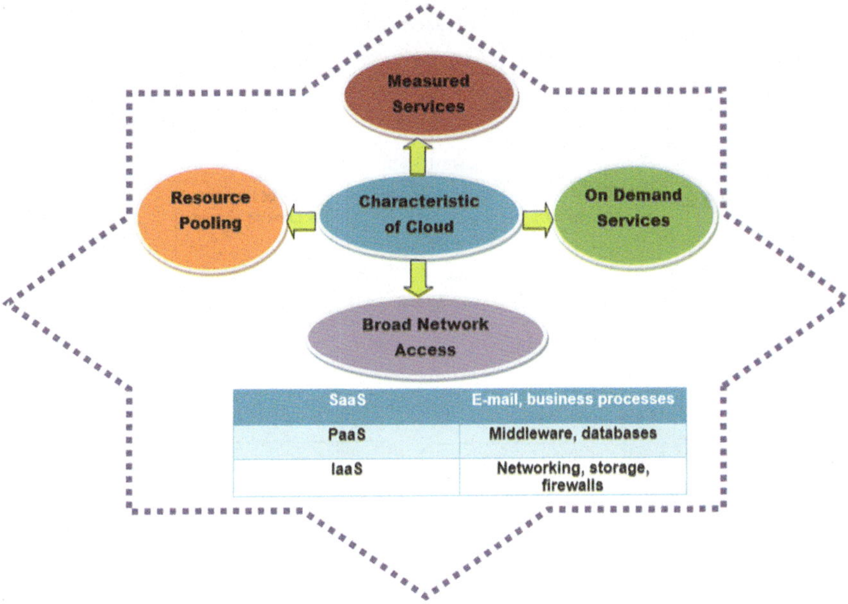

Fig. 2 Cloud paradigm

4 Obstacles in Cloud Computing in IoMT Driven Applications

The cloud computing emerging market carriages many policy-makers' challenges, the IT industry and healthcare groups. E.g., Australia, Europe, Japan and US are observed to be creating strong inroads into cloud computing [22]. On the other hand, important challenges remain i.e. cloud clients require assurances about security and privacy of personal data [23, 24]. A relative survey on Europe, North America, and Asia Pacific by the World Economic Forum and Accenture [25] establish that region of Europe was very concerned with security and privacy issues. Some problems are illustrated in Fig. 3.

5 The IoT Healthcare Network Platform

The IoT healthcare network (IoThNet) platform is an important element of the IoT in health care. It is concerned with the access to the IoT backbone and communication of vital medical data. IoThNet discusses both the computing platform and the network platform model. As presented in Fig. 4, a service platform framework concentrating on residents' health information is offered in [26]. This framework displays a systematic hierarchical model of how agents or caregivers can entrance

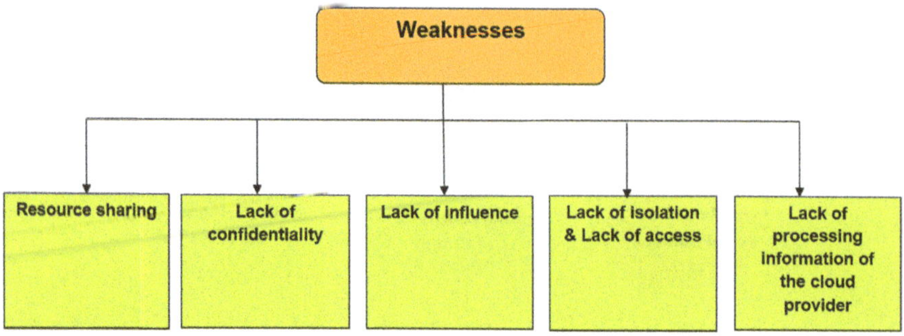

Fig. 3 Weaknesses in cloud computing

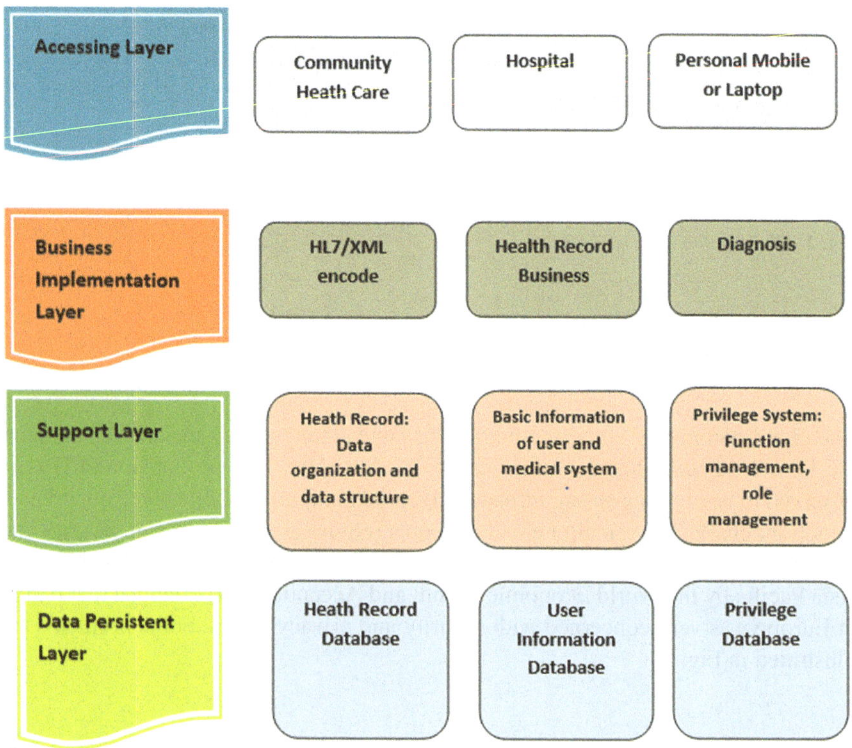

Fig. 4 A health information Service platform framework

different databases from the application layer for the purpose of a support layer. Health record data, basic information of user such as patient, doctor etc., privilege system such as cloud computing and IoT integrated medical function management are included in the support layer. A comparable idea of data center platforms as the middleware between the business layer and smart objects can be detected in [27, 28].

Some experiments [29–31] have been proposed regarding IoThNet platform. Although these research reports have not delivered a generalized and comprehensive analysis including some design issues and models are significant in particular cases. An architecture of semantic platform is presented in [32]. This architecture deals with semantic interoperability across heterogeneous devices and systems and offers demotic devices and user environments with about semantic capability. This semantic layer has four kinds of ontologies.

Correspondingly, a platform is focused on security and privacy issues in medical education learning platform using cloud computing and IoT. It is consisted of two key processes [34], as presented in Fig. 5. The first process mentioned to communication technologies and local storage i.e. data perception, data preprocessing and

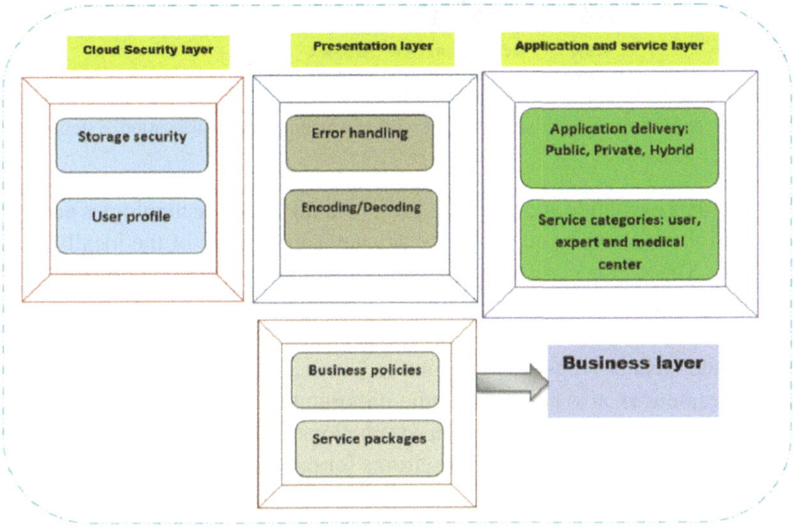

(a) Cloud storage platform

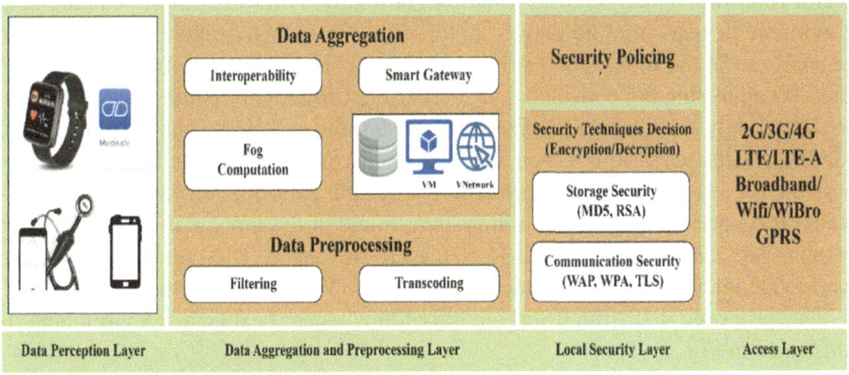

(b) Local storage platform

Fig. 5 IoTHeF cloud storage platforms [6, 33]

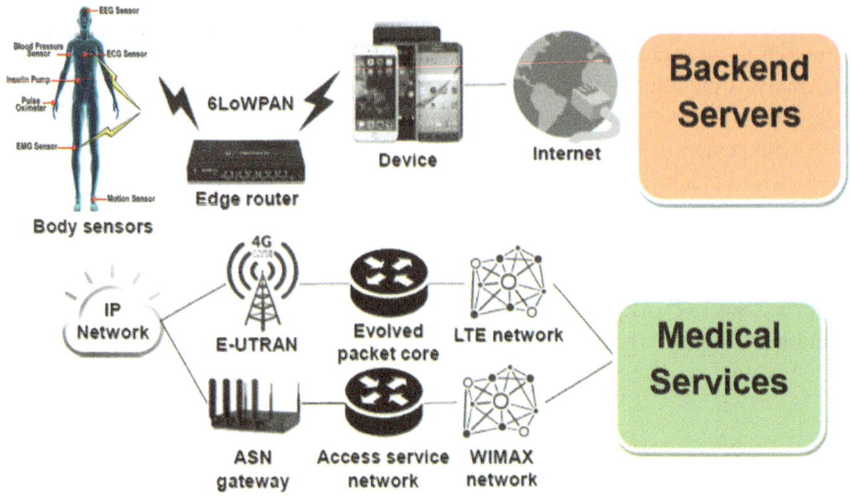

Fig. 6 Remote patient monitoring arrangement with two standard communication systems

aggregation (data exchange and preprocessing of data), access technologies and local security (various measures to confirm the privacy and security of the locally stored patient data). The second process achieved communication and cloud storage i.e. cloud security (cloud storage security and identification of user), presentation (error handling, data decoding, data encoding), service and application (application transfer processes), and business (business policies and service packages [35]. Furthermore, the offered framework also presented some important functions i.e. continuous data exchange between various modules. These functions confirmed that the framework delivered the best privacy and security settings for patients' healthcare data.

Figure 6 illustrates a simple remote patient monitoring arrangement. In this arrangement, the sensors attached in the patient's body can collect the patient's health condition and crucial information constantly. After that, these data are sent to portable devices (mobile phone, laptop etc.) via an edge router. These data will be analyzed in the router and stored on a cloud computing platform for evaluation later. By analyzing the collected data, caregivers can monitor patients remotely and provide timely treatment when their health statuses reveal that they are in critical condition. This scenario is a typical application scenario for IoT in health care. It is possible for different healthcare systems to exchange information with fourth generation (4G) wireless communication system. Some of these popular communication systems are long-term evolution (LTE), LTE Advanced, LTE Advanced Pro and worldwide interoperability for microwave access (WiMAX) [36–40] systems.

The proposed IoT framework in providing smart healthcare environment is illustrated in Fig. 7.

The proposed framework has five layers as shown in Fig. 7. These are presentation, integration, layer of services, analytical data layer, and data layers. In the presentation layer, there are devices with sensors and Internet connection such as smart

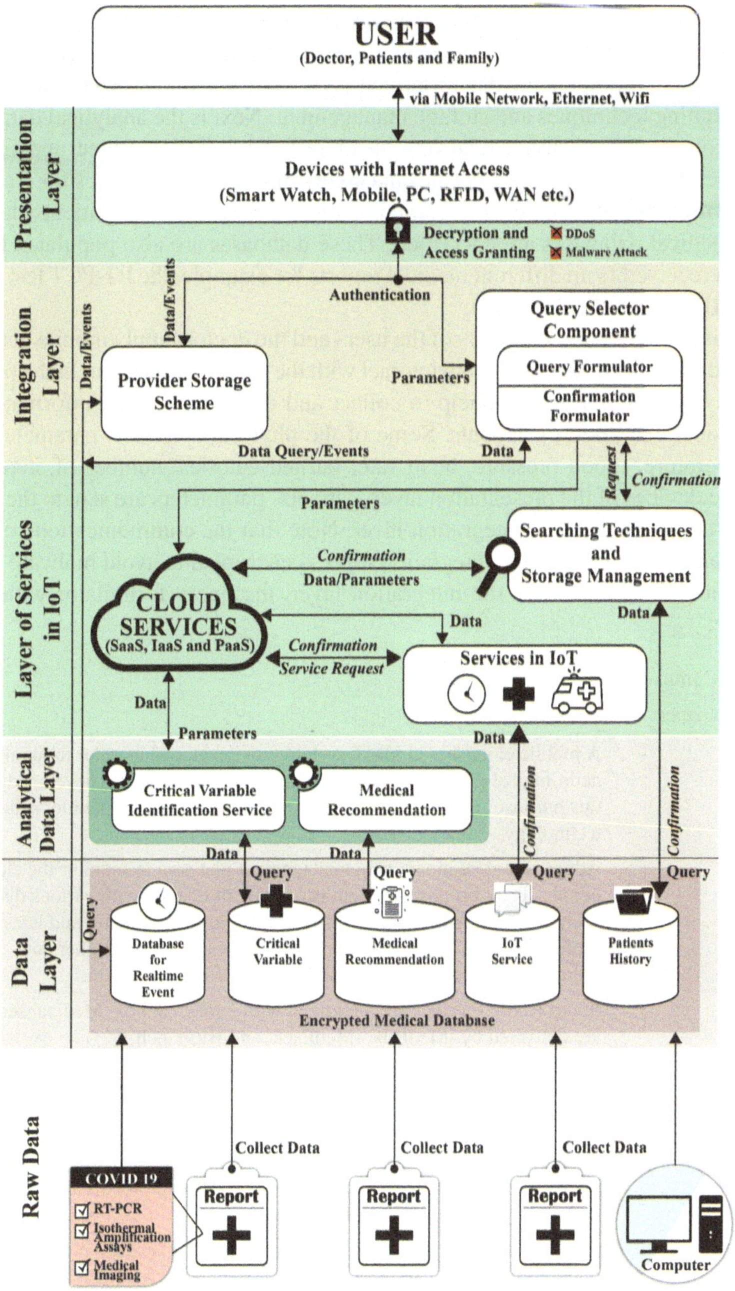

Fig. 7 IoMT framework

phone, smart watch, tablet computer, RFID, etc. Next is the integration layer where the main components are the provider storage scheme and the query selector component. The main components of the service layer are the cloud services, services in IoT, searching techniques and storage management. Next is the analytical data layer having components named critical variable identification (CVI) service, and medical recommendation. The last layer is the data layer consisting of databases for real-time event, critical variable, medical recommendation, IoT service and patients' history. These medical databases are encrypted. These databases are also populated by the raw data received from different medical reports for example, the RT-PCT test report of COVID-19 patients (Table 1).

The users, the family members of the users and the doctors apply mobile network, WiFi and Ethernet based Internet to interact with the presentation layer. In the presentation layer, wearable sensors help to collect and then transmit the information of the physical condition of patients. Some of the physical condition parameters can be temperature, blood pressure, heart rate, burned calories, number of steps, etc. From the devices of the presentation layer, patients' parameters are sent to the query selector component in the integration layer. Note that the communication between the presentation layer and the integration layer is encrypted to avoid malware attack as well as DDoS attack. In the integration layer, the query formulator within the

Table 1 Comparison with other models

Research paper	Features
[41]	A mobile cloud-based IoMT system was developed in order to monitor the neurological disordered patient. Android application was developed for this purpose. They also integrated their system with simple blockchain technology.
[42]	Adopted remote monitoring management platform to monitor the elderly people. Their proposed system is capable of dealing with various disease such as kidney, stroke, heart disease. But uniform standard and latest protocols need to be formulated in body area network. Higher data security system is not maintained in their proposed system.
[43]	Smart health services are integrated with a medical box. Medical services are delivered by iMedBox, iMedPack, and Bio-Patch.
[44]	A smart healthcare system is proposed using a lightweight homomorphism algorithm along with a modified DES algorithm.
[45]	This framework consists of three major layers. One main component of the framework is the IoT gateway. Overall the model focuses on the interoperability, standardization, communication protocols and Internet technology aspects of smart healthcare.
Our proposed work	Our proposed framework has five major layers. These are presentation, integration, layer of services, analytical data layer, and data layers. The communication between presentation and integration layer is encrypted to avoid malware and DDoS attacks. There service layers has one important unit named cloud services which can be of SaaS, IaaS or PaaS types. The data layer can directly populated by raw data for example from RT-PCR test reports of COVID-19 patients.

query selector component requests information from the medical databasc of data layer. For this the query selector unit communicates with the searching technique and storage management unit of service layer who can collect the information directly from the data layer. The searching technique unit can also communicate with the cloud services where the cloud services have direct link with the IoT services. The IoT services unit in the service layer has direct access to the IoT database in the data layer bypassing the analytical data layer. Moreover, the cloud services communicate with the CVI service positioned in the analytical data layer. The CVI service collects the patients' information available in the critical database of data layer. After consultation with the data layer, the CVI layer analyzes the information of patients in order to find out whether the parameters are within the reference range or not. This enables the CVI service to detect a possible emergency condition of patients. In case of emergency, the CVI service informs the cloud service which in turn communicates with the services in IoT unit to request an emergency service, for example, ambulance service. In this process, IoT unit also sends the service provider the medical parameters and current location of the patients, and sends this information to the family members and the doctors.

6 IoT Healthcare Services and Applications

IoMT is applied in various fields of health care, including care for pediatric and elderly patients, automated diagnosis of cancer patients such as lung nodule detection, breast cancer detection, the supervision of chronic diseases, and the management of private health and fitness, among others. In this chapter, the discussion is broadly divided in two sections for better understanding. These two sections are services and applications. Applications are further divided into two groups: single- and clustered-condition applications. A single-condition application means an application that is for a particular disease. On the contrary, a clustered-condition application refers to the case where there are multiple diseases to deal with IoT healthcare services. Figure 8 illustrates this categorization.

Single condition applications of IoMT include:

(a) Sensing the glucose level in order to manage the diabetes patients
(b) Monitoring the Electrocardiogram (ECG) in order to measure the heart rate and rhythm
(c) Blood pressure monitoring for pressure sensitive patients
(d) Body temperature monitoring for viral fever or fever from cold
(e) Oxygen saturation monitoring

Clustered condition applications of IoMT include:

(a) Rehabilitation management
(b) Medication management
(c) Wheelchair management

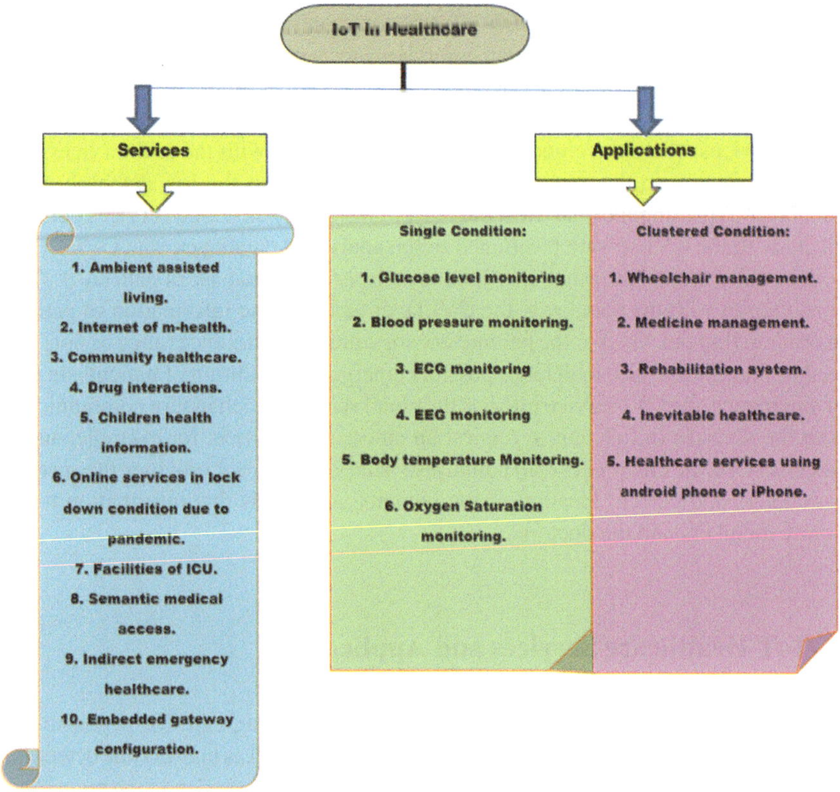

Fig. 8 IoT healthcare services and applications

(d) Imminent healthcare
(e) Healthcare solution by android or i-phone apps

A number of healthcare applications have been reported in recent years. The applications of IoT in healthcare sector largely depend on the use of sensors. This is because medical sensors capture the vital information of patients and then transmit the information via network. The use of sensors allow doctors to monitor the real-time condition of patients and thus to provide necessary treatment. The sensors are capable of evaluating different health related information for example, blood pressure, arterial oxygen, emotion tracking, blood glucose level, etc. The sleep quality of patients can also be tracked by the use of sensors.

7 Classification of Healthcare Apps by Category

Various hardware and software products have been designed to make smartphones a versatile healthcare device. There are many healthcare smartphone applications (apps) suitable for different scenarios. Since there are significant development in the data-driven diagnosis of diseases [46, 47], the apps can play a vital role in disease diagnosis. The work in [48] provides a review of such apps. Some of these apps include patients and general healthcare apps, medical education, training, information search apps. These apps are in generally called auxiliary apps. In addition, there are many recent apps serving similar purposes [49–55]. Based on these references, Fig. 9 presents a classification diagram of auxiliary apps. Table 2 summarizes the various sensors used in specific disease with the process of transmitting data.Security must be ensured so that malware or ransomware cannot be attacked inside the user's phone at the time of downloading or installing the apps from play store or other repositories [74, 75].

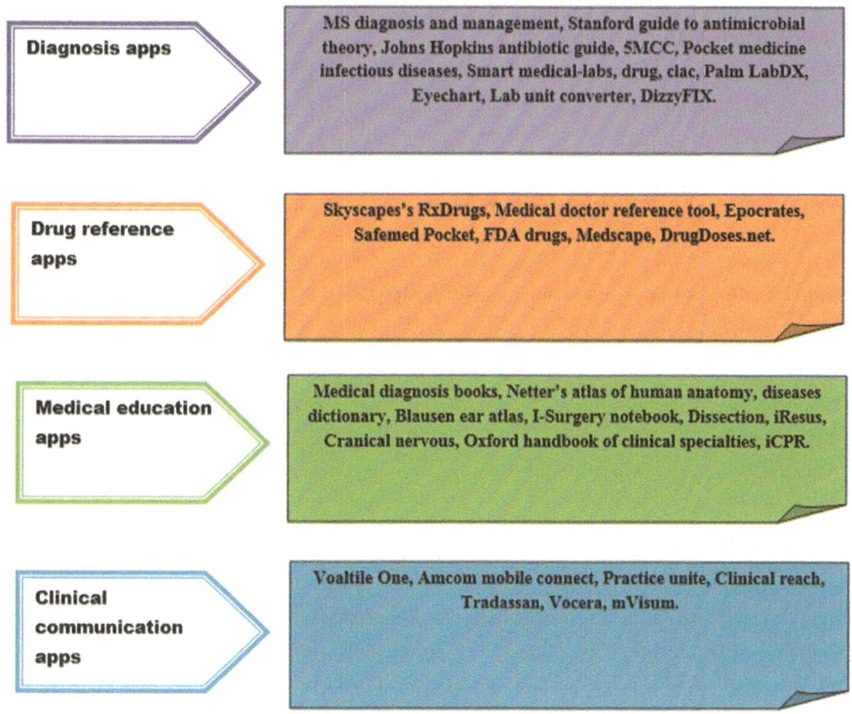

Fig. 9 Auxiliary apps in IoMT healthcare for smartphones

Table 2 Various sensors used in specific disease with the process of transmitting data

Disease	Sensors	Role of IoT
Diabetes [56, 57]	Glucose sensor, contextual sensor, NEAR IR LED	Firstly, the output from the desired sensor is sent to the Android gateway for local storage and pre-processing of the data. After that operation, the cloud receives the data for analysis and notifies the responsible person when any abnormalities is found.
Asthma [58, 59]	Pulse sensor, temperature sensor	MC board processes the necessary signals. Then data is sent with the help of Wi-Fi to the cloud for data storage via HTTP protocol.
Heart disease [48, 60–66]	Optical heart rate sensor, blood pressure (BP) sensor, electrocardiogram (ECG) sensor	Wireless transmitter transmits the data to a microcontroller. After that, it is sent to the servers through a smartphone gateway.
Hyperthermia and hypothermia [67–69]	Thermopile IR sensor, wearable thermometry	Wireless body area network (WBAN) connects the sensors via gateway, then the raw data is sent to the server through Wi-Fi communication for storage and analyzing.
Wheelchair management [70, 71]	Camera sensor, force sensor, accelerometer sensor	The controller is integrated into the wheelchair processed signals from sensors and therefore it can realize the abnormality condition.
Neuromuscular diseases [72, 73]	Electromyography (EMG) sensors	A signal classification model is applied in order to detect the neuromuscular disorders after sending the raw data to the controller. Then the classification results are sent to the cloud for storage.

8 Conclusion

This chapter provides a comprehensive description of the strengths, weakness, prospects and challenges of the IoMT integrated cloud computing for healthcare system. In an IoMT scheme, sensors around patients capture data and send to the cloud for measurement, analysis, making decisions and taking action plans. Apart from sensor technology, artificial intelligence, big data, data analytics are driving the progress of IoMT. The success of IoMT has been possible because of the introduction of smartphones with near field communication and radio frequency identification tags, mobile applications, medical devices, Internet connectivity, modern hospital infrastructure, etc. With the continuous development of IoMT, the traditional healthcare scheme is expected to move to a proactive patient care system. However, security and privacy concerns are still to overcome for the widespread adoption of IoMT using the cloud. In future software and hardware tools should be used to ensure security of IoMT as the impact of any security breach of data is significant. With the current spread of 'severe acute respiratory syndrome coronavirus 2′ (SARS-CoV-2) that causes coronavirus disease 2019 (COVID-19), the importance of IoMT integrated cloud is now higher than ever. Hence, more studies are required to make IoMT integrated cloud computing available to everyone everywhere in the world.

References

1. Höller, J., Tsiatsis, V., Mulligan, C., Karnouskos, S., Avesand, S., Boyle, D.: From Machine-to Machine to the Internet of Things: Introduction to a New Age of Intelligence. Elsevier, Amsterdam (2014)
2. Bharati, S., Podder, P., Mondal, M.R.H., Robel, M.R.A.: Threats and countermeasures of cyber security in direct and remote vehicle communication systems. J. Inf. Assur. Secur. **15**(4), 153–164 (2020)
3. Islam, S.M.R., Kwak, D., Kabir, M.H., Hossain, M., Kwak, K.: The Internet of Things for health care: a comprehensive survey. IEEE Access **3**, 678–708 (2015)
4. Pang, Z.: Technologies and architectures of the Internet-of-Things (IoT) for health and well-being. M.S. thesis, Depart. Electron. Comput. Syst., KTH-Roy. Inst. Technol., Stockholm, Sweden, Jan (2013)
5. Truong, H.L., Dustdar, S.: Principles for engineering IoT cloud systems. IEEE Cloud Comput. **2**, 68–76 (2015)
6. Robel M.R.A., Bharati S., Podder P., Raihan-Al-Masud M., Mandal S.: Fault tolerance in cloud computing- an algorithmic approach. In: Abraham, A., Panda, M., Pradhan, S., Garcia-Hernandez, L., Ma, K. (eds.) Innovations in Bio-Inspired Computing and Applications. IBICA 2019. Advances in Intelligent Systems and Computing, vol 1180. Springer, Cham (2021). https://doi.org/10.1007/978-3-030-49339-4_31
7. Internet of Things at a Glance: Available online: https://www.cisco.com/c/dam/en/us/products/collateral/se/internet-of-things/at-a-glance-c45-731471.pdf. Accessed 23 Feb 2019
8. Size of the Internet of Things MarketWorldwide in 2014 and 2020, by Industry. Available online: https://www.statista.com/statistics/512673/worldwide-internet-of-things-market/. Accessed on 24 Feb 2019
9. Borgia, E.: The internet of things vision: key features, applications, and open issues. Comput. Commun. **54**, 1–31 (2014)
10. Bhushan, B., Khamparia, A., Sagayam, K. M., Sharma, S. K., Ahad, M. A., and Debnath, N. C.: Blockchain for smart cities: a review of architectures, integration trends and future research directions. Sustainable. Cities. Soc. **61**:102360 (2020). https://doi.org/10.1016/j.scs.2020.102360
11. Mineraud, J., Mazhelis, O., Su, X., Tarkoma, S.: A gap analysis of internet-of-things platforms. Comput. Commun. (2016). https://doi.org/10.1016/j.comcom.2016.03.015
12. Muhammad, K., Ahmad, J., Rho, S., Baik, S.: Image steganography for authenticity of visual contents in social networks. Multimedia Tools Appl. 1–20 (2017)
13. Muhammad, K., Sajjad, M., Lee, M. Y., Baik, S.W.: Efficient visual attention driven framework for key frames extraction from hysteroscopy videos. Biomed. Signal Process. Control. **33**, 161–168 (2017)
14. Khamparia, A., Gupta, D., de Albuquerque, V.H.C., et al.: Internet of health things-driven deep learning system for detection and classification of cervical cells using transfer learning. J. Supercomput. (2020). https://doi.org/10.1007/s11227-020-03159-4

15. Bharati, S., Podder, P., Mondal, M.R.H.: Hybrid deep learning for detecting lung diseases from X-ray images. Inform. Med. Unlocked, **20**:100391 (2020). https://doi.org/10.1016/j.imu.2020.100391
16. Shahariar Parvez, A.H.M., Robiul Alam Robel, M., Rouf, M.A., Podder, P., Bharati, S.: Effect of fault tolerance in the field of cloud computing. In: Smys, S., Bestak, R., Rocha, Á. (eds.) Inventive Computation Technologies. ICICIT 2019. Lecture Notes in Networks and Systems, vol. 98. Springer, Cham (2020)
17. Katz, R., Goldstein, P., Yanosky, R.: Cloud computing in higher. EDUCAUSE (2010). http://net.educause.edu/sectionparams/conf/CCW10/highered.pdf
18. Chang, V.: An overview, examples, and impacts offered by emerging services and analytics in cloud computing virtual reality. Neural Comput. Appl. **29**, 1243–1256 (2018). https://doi.org/10.1007/s00521-017-3000-1
19. Buyya, R., Yeo, C., Venugopal, S., Broberg, J., Brandic, I.: Cloud computing and emerging IT platforms: vision, hype, and reality for delivering computing as the 5th utility. Fut. Gener. Comput. Syst. **25**(6), 559–616 (2009)
20. Granell, C., Havlik, D., Schade, S., Sabeur, Z., Delaney, C., Pielorz, J., Thomas, U., Mazzetti, P., Schleidt, K., Kobernus, M., Havlik, F., Bodsberg, N., Berre, A., Lorenzo, J.: Future internet technologies for environmental applications. Environ. Model. Softw. **15**(78) (2016)
21. Flammini, A., Sisinni, E.: Wireless sensor networking in the internet of things and cloud computing era. Proced. Eng. **87**, 672–679 (2014)
22. Martin, R.: Japan is best prepared to capitalize on cloud computing (2012). http://www.techinasia.com/japan-cloud-cloud-computing/
23. Venters, W.: A critical review of cloud computing: researching desires and realities. J. Inf. Technol. **27**, 79–97 (2012)
24. Gutub, A., Alharth, N.: Improving Hajj and Umrah services utilizing exploratory data visualization techniques. Hajj forum 2016—the 16th scientific Hajj research Forum, organized by the custodian of the two holy Mosques Institute for Hajj Research. Umm Al-Qura University—King Abdulaziz Historical Hall, Makkah, pp. 561–572 (2016)
25. Berry, R., Reisman, M.: Policy challenges of cross-border cloud computing. J. Int. Commer. Econ. **4**(2), 1–38 (2012)
26. Wang, W., Li, J., Wang, L., Zhao, W.: The Internet of Things for resident health information service platform research. In: Proceedings of IET International Conference Communication Technology Applications (ICCTA), Oct 2011, pp. 631–635
27. Yang, L., Ge, Y., Li, W., Rao, W., Shen, W.: A home mobile healthcare system for wheelchair users. In: Proceedings of IEEE International Conference Computer Supported Cooperative Work Design (CSCWD), May 2014, pp. 609–614
28. Darwish, A., Hassanien, A.E., Elhoseny, M., et al.: The impact of the hybrid platform of internet of things and cloud computing on healthcare systems: opportunities, challenges, and open problems. J. Ambient Intell. Human. Comput. **10**, 4151–4166 (2019). https://doi.org/10.1007/s12652-017-0659-1
29. Jara, A.J., Zamora-Izquierdo, M.A., Skarmeta, A.F.: Interconnection framework for mHealth and remote monitoring based on the Internet of Things. IEEE J. Sel. Areas Commun. **31**(9), 47–65 (2013)
30. Bharati, S., Podder, P. and Mondal, M.R.H.: Artificial neural network based breast cancer screening: a comprehensive review. Int. J. Comput. Inf. Syst. Ind. Manag. Appl. **12**, 125–137 (2020).
31. Jia, X., Chen, H., Qi, F.: Technical models and key technologies of e-health monitoring. In: Proceedings of IEEE International Conference on e-Health Network, Applications and Services (Healthcom), Oct 2012, pp. 23–26
32. Miori, V., Russo, D.: Anticipating health hazards through an ontology-based, IoT domotic environment. In: Proceedings of 6th International Conference on Innov. Mobile Internet Services Ubiquitous Computing (IMIS), July 2012, pp. 745–750
33. Ali, M., Bilal, H.S.M., Razzaq, M.A., Khan, J., Lee, S., Idris, M., Aazam, M., Choi, T., Han, S.C., Kang, B.H.: IoTFLiP: IoT-based flipped learning platform for medical education. Digit. Commun. Netw. **3**, 188–194 (2017)

34. Podder, P., Bharati, S., Robel, M.R.A., Raihan-Al-Masud, M., Rahman, M.A.: Uplink and downlink spectral efficiency estimation for multi antenna MIMO user. In: Abraham, A., Panda, M., Pradhan, S., Garcia-Hernandez, L., Ma, K. (eds.) Innovations in Bio-Inspired Computing and Applications. IBICA 2019. Advances in Intelligent Systems and Computing, vol. 1180. Springer, Cham (2021). https://doi.org/10.1007/978-3-030-49339-4_29
35. Verma, P., Sood, S.K., Kalra, S.: Cloud-centric IoT based student healthcare monitoring framework. J. Ambient Intell. Human. Comput. **9**, 1293–1309 (2018). https://doi.org/10.1007/s12652-017-0520-6
36. Abideen, Z.U.; Shah, M.A. (2017). An IoT based robust healthcare model for continuous health monitoring. In: Proceedings of the 23rd International Conference on Automation and Computing (ICAC), Huddersfield, UK, 7–8 Sept 2017, pp. 1–6
37. Sarker, N., Islam, M.A., Mondal, M.R.H.: Two novel multiband centimetre-wave patch antennas for a novel OFDM based RFID system. J. Commun. (JCM) **13**(6) (2018)
38. Bharati, S., Podder, P.: Adaptive PAPR reduction scheme for OFDM using SLM with the fusion of proposed clipping and filtering technique in order to diminish PAPR and signal distortion. Wirel. Pers. Commun. **113**, 2271–2288 (2020). https://doi.org/10.1007/s11277-020-07323-0
39. Rahmani, A.M., Gia, T.N., Negash, B., Anzanpour, A., Azimi, I., Jiang, M., Liljeberg, P.: Exploiting smart e-Health gateways at the edge of healthcare Internet-of-Things: a fog computing approach. Fut. Gener. Comput. Syst. **78**, 641–658 (2018)
40. Luhach, A.K., Khamparia, A., Sihag, R., Kumar, R.: Honey bee optimization based sink mobility aware heterogeneous protocol for wireless sensor network. Scalable Comput. Pract. Exp. **20**(4), 591–598, (2020)
41. Nguyen, D.C., Nguyen, K.D., Pathirana, P.N.: A mobile cloud based IoMT framework for automated health assessment and management. In: 2019 41st Annual International Conference of the IEEE Engineering in Medicine and Biology Society (EMBC), Berlin, Germany, pp. 6517–6520 (2019). 10.1109/EMBC.2019.8856631
42. Zhao, W., Wang, C., Nakahira, Y.: Medical application on internet of things. In: Proceedings of IET International Conference on Communication Technology and Application (ICCTA 2011), pp. 660–665, Beijing, China (2011)
43. Yang, G., Xie, L., Mäntysalo, M., et al.: A health-IoT platform based on the integration of intelligent packaging, unobtrusive bio-sensor, and intelligent medicine box. IEEE Trans. Ind. Inf. **10**(4), 2180–2191 (2014)
44. Gong, T., Huang, H., Li, P., Zhang, K., Jiang, H.: A medical healthcare system for privacy protection based on IoT. In: Proceedings of the Parallel Architectures, Algorithms and Programming (PAAP), 2015 Seventh International Symposium on, pp. 217–222, Dec 2015
45. Pasha, M., Shah, S.M.W.: Framework for E-health systems in IoT-based environments. Wireless Commun. Mobile Comput. **2018**. Article ID 6183732 (2018). https://doi.org/10.1155/2018/6183732
46. Raihan-Al-Masud, M., Mondal, M.R.H.: Data-driven diagnosis of spinal abnormalities using feature selection and machine learning algorithms. PLoS ONE **15**(2), e0228422 (2020)
47. Bharati, S., Podder, P., Mondal, R., Mahmood, A., Raihan-Al-Masud, M.: Comparative performance analysis of different classification algorithm for the purpose of prediction of lung cancer. In: Abraham, A., Cherukuri, A., Melin, P., Gandhi, N. (eds.) Intelligent Systems Design and Applications. Advances in Intelligent Systems and Computing, vol. 941, pp. 447–457 (2018)
48. Mosa, A.S.M., Yoo, I., Sheets, L.: A systematic review of healthcare applications for smartphones. BMC Med. Inf. Des. Making **12**, 67 (2012)
49. [Online]. Available: http://www.medicaljoyworks.com. Last accessed on 20 Apr 2020
50. [Online]. Available: http://www.prognosisapp.com. Last accessed on 20 Apr 2020
51. [Online]. Available: http://www.imedicalapps.com/2014/01/diagnoseapp-evidence-based-clinical-decision. Last Accessed on 20 Apr 2020
52. [Online]. Available: http://www.eagleget.com/apps/apk-file/4470. Last accessed on 20 Apr 2020
53. [Online]. Available: https://www.apple.com/itunes/charts. Last accessed on 20 Apr 2020
54. [Online]. Available: https://play.google.com/store/apps?hl=en. Last accessed on 20 Apr 2020

55. White, P.J.F., Podaima, B.W., Friesen, M.R.: Algorithms for smartphone and tablet image analysis for healthcare applications. IEEE Access **2**, 831840 (2014)
56. Gia, T.N., Ali, M., Dhaou, I.B., Rahmani, A.M., Westerlund, T., Liljeberg, P., Tenhunen, H.: IoT-based continuous glucose monitoring system: a feasibility study. Procedia Comput. Sci. **109**, 327–334 (2017)
57. Sunny, S.; Kumar, S.S. Optical based non invasive glucometer with IoT. In: Proceedings of the 2018 International Conference on Power, Signals, Control and Computation (EPSCICON), Thrissur, India, 6–10 Jan 2018; pp. 1–3
58. AL-Jaf, T.G., Al-Hemiary, E.H.: Internet of Things based cloud smart monitoring for asthma patient. In: Proceedings of the 1st International Conference on Information Technology (ICoIT'17), Erbil, Iraq, 10 Apr 2017, p. 380
59. Raji, A., Devi, P.K., Jeyaseeli, P.G., Balaganesh, N.: Respiratory monitoring system for asthma patients based on IoT. In: Proceedings of the 2016 Online International Conference on Green Engineering and Technologies (IC-GET), Coimbatore, India, 19 Nov 2016; pp. 1–6
60. Satija, U., Ramkumar, B., Manikandan, M.S.: Real-time signal quality-aware ECG telemetry system for IoT-based health care monitoring. IEEE Internet Things J. **4**, 815–823 (2017)
61. Beach, C., Krachunov, S., Pope, J., Fafoutis, X., Piechocki, R.J., Craddock, I., Casson, A.J.: An ultra low power personalizable wrist worn ECG monitor integrated with IoT infrastructure. IEEE Access **6**, 44010–44021 (2018)
62. Sobya, D., Muruganandham, S., Nallusamy, S., Chakraborty, P.: Wireless ECG monitoring system using IoT based signal conditioning module for real time signal acquisition. Indian J. Publ. Health Res. Dev. **9**, 294–299 (2018)
63. He, J., Rong, J., Sun, L., Wang, H., Zhang, Y., Ma, J.: D-ECG: a dynamic framework for cardiac arrhythmia detection from IoT-based ECGs. In: Proceedings of the International Conference on Web Information Systems Engineering, Dubai, UAE, 12–15 Nov 2018; pp. 85–99
64. Bansal, M., Gandhi, B.: IoT based smart health care system using CNT electrodes (for continuous ECG monitoring). In: Proceedings of the 2017 International Conference on Computing, Communication and Automation (ICCCA), Greater Noida, India, 5–6 May 2017, pp. 1324–1329
65. Xin, Q., Wu, J.: A novel wearable device for continuous, non-invasion blood pressure measurement. Comput. Biol. Chem. **69**, 134–137 (2017)
66. Chao, P.C.P., Tu, T.Y.: Using the time-domain characterization for estimation continuous blood pressure via neural network method. In: ASME 2017 Conference on Information Storage and Processing Systems collocated with the ASME 2017 Conference on Information Storage and Processing Systems, San Francisco, CA, USA, 29–30 Aug 2017; p. V001T02A003
67. Huang, M., Tamura, T., Tang, Z., Chen, W., Kanaya, S.: A Wearable thermometry for core body temperature measurement and its experimental verification. IEEE J. Biomed. Health Inform. **21**, 708–714 (2017)
68. Li, Q., Zhang, L.N., Tao, X.M., Ding, X.: Review of flexible temperature sensing networks for wearable physiological monitoring. Adv. Healthc. Mater. **6**, 1601371 (2017)
69. Ota, H., Chao, M., Gao, Y., Wu, E., Tai, L.C., Chen, K., Matsuoka, Y., Iwai, K., Fahad, H.M., Gao, W., et al.: 3d printed "earable" smart devices for real-time detection of core body temperature. ACS Sens. **2**, 990–997 (2017)
70. Ghorbel, A., Bouguerra, S., Amor, N.B., Jallouli, M.: Cloud based mobile application for remote control of intelligent wheelchair. In: Proceedings of the 2018 14th International Wireless Communications & Mobile Computing Conference (IWCMC), Limassol, Cyprus, 25–29 June 2018; pp. 1249–1254
71. Lee, Y.K.; Lim, J.M.; Eu, K.S.; Goh, Y.H.; Tew, Y. Real time image processing based obstacle avoidance and navigation system for autonomous wheelchair application. In: Proceedings of the Asia-Pacific Signal and Information Processing Association Annual Summit and Conference (APSIPA ASC), Kuala Lumpur, Malaysia, 12–15 Dec 2017; pp. 380–385
72. Subhash, K., Pournami, P., Joseph, P.K.: Census transform based feature extraction of EMG signals for neuromuscular disease classification. In: Proceedings of the 2017 IEEE 15th Student Conference on Research and Development (SCOReD), Putrajaya, Malaysia, 13–14 Dec 2017; pp. 499–503

73. Sadiq Iqbal, M., Nasim Akhtar, M., Shahariar Parvez, A.H.M., Bharati, S., Podder, P.: Ensemble learning-based EEG feature vector analysis for brain computer interface. In: Suma, V., Bouhmala, N., Wang, H. (eds.) Evolutionary Computing and Mobile Sustainable Networks. Lecture Notes on Data Engineering and Communications Technologies, vol. 53. Springer, Singapore (2021). https://doi.org/10.1007/978-981-15-5258-8_88
74. Wazid, M., Das, A.K., Rodrigues, J.J.P.C., Shetty, S., Park, Y.: IoMT malware detection approaches: analysis and research challenges. IEEE Access **7**, 182459–182476 (2019)
75. Podder, P., Mondal, M.R.H., Bharati, B., Paul, P.K.: Review on the security threats of internet of things. Int. J. Comput. Appl. **176** (41), 37–45 (2020). https://doi.org/10.5120/ijca2020920548

Optimal SVM Based Brain Tumor MRI Image Classification in Cloud Internet of Medical Things

S. Chidambaranathan, A. Radhika, Veeraraghavan Vishnu Priya, Surapaneni Krishna Mohan, and M. G. Gireeshan

Abstract At recent days, massive increase in the Internet of Things (IoT) and cloud computing receives numerous healthcare services to the subsequent stage. On the other hand, brain tumor (BT) is recognized as a deadly disease that enhances the annual mortality rate. This study projects a new detection and diagnosis model for BT. Here, two main stages are involved namely feature selection and classification. Initially, examination of the patients takes place using medical devices linked to IoT. When the MRI images of a person are acquired, pre-processing will takes place. Next, improved gravitational search algorithm with genetic algorithm (IGSAGA) model is applied for filtering the features and optimal support vector machine (SVM) model is applied for classification processes. The results are validated using a benchmark BRATS dataset and the experimental outcome indicated the supremacy of the projected model.

S. Chidambaranathan (✉)
Computer Applications, St. Xavier's College Autonomous, Palayamkottai, India
e-mail: scharan2009@gmail.com

A. Radhika
Department of Electrical and Electronics Engineering, Sri Krishna College of Engineering and Technology, Coimbatore, India
e-mail: radhika@skcet.ac.in

V. V. Priya
Department of Biochemistry, Saveetha Dental College, Saveetha Institute of Medical and Technical Sciences, Chennai, India
e-mail: drvishnupriyav@gmail.com

S. K. Mohan
Department of Biochemistry, Panimalar Medical College Hospital & Research Institute, Varadharajapuram, Poonamallee, Chennai, India
e-mail: krishnamohan.surapaneni@gmail.com

M. G. Gireeshan
Jaibharath Arts and Science College, Vengola, Kerala, India
e-mail: Mggireeshan@gmail.com

A. E. Hassanien et al. (eds.), *Cognitive Internet of Medical Things for Smart Healthcare*, Studies in Systems, Decision and Control 311,
https://doi.org/10.1007/978-3-030-55833-8_5

Keywords IoMT · Brain tumor · Classification · Gravitational search algorithm · SVM

1 Introduction

Presently, Internet of Things (IoT) enables the human being, object and virtual environment to interact among one another in a proper way [1]. Diverse applications utilize IoT in data acquisition task for smart platforms like transportation, smart homes, healthcare, etc. Due to the development of the IoT and sensing components, numerous studies have been made [2]. The increase of exclusive medication and existence of different disease in a global way, it mainly requires the growth of healthcare to a maximum extent. For the concentration on the procurer of disease management, it is required to design a model which utilizes the everywhere sensing capabilities of IoT devices to predict the probability of deadly disease exist in the patient.

IoT and cloud computing are interactive things and the combination finds useful to observe the patients exist in the inaccessible regions through providing continuous support from physicians [3]. IoT is managed using virtual unimpeded capabilities and cloud sources for balancing the technological restrictions such as storage, processing and energy. Simultaneously, the cloud provides benefits from IoT through the expansion of its scope to deal with real things in the real world and offered multiple services in a distributed as well as dynamical manner. Therefore, it is needed to develop a new IoT with cloud model for novel applications and services in medicinal field [4, 5].

Internet of Medical Things (IoMT) combines two domains of IoT as well as healthcare [6]. Brain tumor (BT) injures the human at every stage and highly raises the death [7]. A tumor contains several tissues from aggregated abnormal cells. Benign BT is non-not cancerous and unrestrictive. It will not distribute to nearby cells, it is severe in rate cases. The Malignant BT denotes the cancer which generates in the brain and spreads to other regions in a quick way over benign type. Magnetic resonance imaging (MRI) can be applied to quickly understand the details related to tumor and evaluate the distribution rate. Numerous MRI scan images are required to classify it using machine learning (ML) based classifier model. For optimal classification model design, the characteristics which need to be assumed are detection rate and algorithm complexity [8].

By the utilization of unsupervised classifier models such as FCM and Self Organizing Map (SOM) and supervised models like k-NN, support vector machine (SVM) and artificial neural network (ANN), the brain MRI images will be classified. Generative and discriminative models are two types of automated segmentation model for BT. When compared to independent models [9], present studies indicated that the approaches depend upon discriminative classification indicates better outcome. The relativity between the ground truth and input image are learned through discriminative approaches [10]. Under the valid ground truth rates, supervised learning model is applied in various scenarios which need large dataset. By the application of earlier knowledge like spatial extent of healthy cells and position, the probability based models are created by generative approaches. For deriving the unknown tumor parts, priory gained copy of healthy tissues is applied. However, it is a hard task of converting the earlier information to proper probability based model [11–14].

This study projects a new detection and diagnosis model for BT. Here, two main stages are involved namely feature selection and classification. Initially, examination of the patients takes place using medical devices linked to IoT. When the MRI images of a person are acquired, pre-processing will takes place. Next, improved gravitational search algorithm with genetic algorithm (IGSAGA) model is applied for filtering the features and optimal support vector machine (SVM) model is applied for classification processes. The results are validated using a benchmark BRATS dataset and the experimental outcome indicated the supremacy of the projected model.

2 Background Information

IoT biological system has an unpredictable architecture, in which different segments interact with each other to enable various answers for the end client. This is a reliant framework, which enables real-time data acquisition, gadget network, data transfers, and analytics to control end client applications. IoT gives the associated condition, containing the digital physical frameworks, which integrates human mediation with PC based frameworks and facilitates data-driven choice procedures. Right now, IoT encompasses advances, for example, smart lattices, smart homes, keen coordinations, and smart towns, augmented through sensor, actuator, and communication convention systems. IoT offers various real-time arrangements through the integration of data analytics and sensors implanted on machines. There is a noteworthy move in the innovation division that has let everybody stunned alone for their brains. A few has figured out how to remain side by side with it while others are attempting to get a hold tight the things. In any case, regardless of the amount it develops, there will consistently be need of manual basic leadership capacity. In a course of time, there is more effectiveness in the work procedure and errand the board.

The main three words that can characterize robotization, comfort, and effectiveness, at the same time is IoT. It has changed the course of numerous businesses making a ceaseless buzz. This innovation permits associating gadgets with the assistance of switch—on and off—and screens information which is additionally associated with the cloud. This has figured out how to impact the social insurance part too. IoMT is utilized for the aggregate gadgets that are utilized in restorative science and applications that help in associating PC coordinate with the human services IT frameworks. In this new innovation, the therapeutic hardware or gadgets are associated with the web innovation, for example, Wi-Fi that enables a simple machine-to-machine correspondence. Likewise, the IoMT gadgets are additionally associated with the cloud-based stages that incorporate Amazon Web Services that stores and break down the caught information. This entire procedure is additionally generally known as social insurance IoT. There is no uncertainty that the therapeutic business has consistently been progressively open to exploring different avenues regarding new innovations and any of headway in the IT division. When IoT was actualized in the Medical business then nobody has imagined that it will turn out to be such a serious piece of innovation.

Turning Data to Actions
The eventual fate of medicinal science has a tight hang on evaluated wellbeing with the goal that the quantifiable medical advantages can be improved. This single explanation is making the certified wellbeing innovation an immense blast in the medicinal science. Likewise, everybody knows the significance of information for a superior result and improved execution. Consequently, the necessity of IoT in drugs is enhanced with wellbeing following framework and estimation by utilizing different items.

Promoting Preventative Care
The human services isn't more centred around avoiding an issue as opposed to restoring. This essential objective is towards the future where the fix or treatment pace of any ailment will be hard to oversee. It gives an entrance to profoundly dependable and ongoing information on the soundness of a person. This opens a colossal open door for the medicinal services to help individuals and guide them to carry on with a solid life.

Advancing Wellbeing
Is it safe to say that it isn't being extraordinary, if the wearable gadgets tend to screen the pulse of a patient and can tell when the perusing isn't on track? Not on this, envision it imparting this data to the next hardware that is utilized for the practicing reason that you use while working out. This will viably take out the necessity of EMRs as the data will straightforwardly store to cloud. The put away data is then utilized by the specialists to think of some treatment choice for a patient. It is an incredible observing apparatus and restorative adherence.

Boosting Fulfilment and Involvement
IoT can advance careful errands that will consequently give back the fulfilled client. Likewise, it is an extraordinary method to get patient associated with every single step of the treatment as opposed to keeping him as an afterthought track. This will keep the patient draw in with their doctors that can help in the decrease of the meeting him by and by. The gadgets will store all the fundamental information that can be utilized by the patient and doctor.

Enhancing Health Control
Through IoT innovation, one can peruse the wearable information perusing and the information is caught on the cloud that can never be passed up a great opportunity. This is typically missed in HER that settle on the errand of settling on the choice somewhat scary. The medicinal services group legitimately gets the subtleties that are drafted by IoT checking of any kind of constant illnesses [15].

Enhancing Care Organization
Through this, the consideration group can without much of a stretch gather and interface the information that is come about by the wearable gadgets, for example, sweat, pulse, movement, rest, and temperature. The sensor-sustained data helps in sending a caution to the individual patient and care supervisory crew continuously as it were. This outcome is getting a caution when an issue emerges. This outcome of this work process streamlining and guarantees that the patient is figured out how to from the bounds of his home.

2.1 Challenges and Advantages of IoMT

The contribution of a savvy coordinate with IoT that will change the fate of the social insurance. A gigantic innovatory change in the advancements of the sensor that will roll out a major improvement later on if IoT. The examination will be progressed soon that will permit passing a lot of information to the IoT gadgets that can be utilized according to the prerequisite [16].

The distributed computing innovation will be increasingly created as far as speed that will effectively help the propelled rendition of IoT innovation.

The work on the security will be the point of the progressed IoT to guarantee the security of the information gathered by patients progressively [17].

Advantages or Favourable Circumstances of IoMT
Following are the advantages or favourable circumstances of IoMT:

- Patients get new treatment offices utilizing better human services gadgets and drugs at reasonable expense. This brings down by and large cost for patients.
- It offers better treatment results.
- It guarantees more trust towards specialists. This is because of the way that IoMT advances support abilities of specialists and scientists.
- Mistakes are diminished to bigger degree.
- Intake of meds can be checked and controlled.
- It is anything but difficult to keep up utilization of restorative gadgets in such IoMT organize.
- It offers better command over illnesses.
- Hospitals can follow patients and medications by getting to information by means of applications introduced on their cell phones from associated gadgets for example glucose meter, pulse meter, etc.

2.2 Technologies Enduing IoMT Implementation

The macro-level IoT architecture includes three layers: local gadgets, availability, and data analytics and arrangements. The structural and functional aspects of each segment are talked about as pursues:

2.2.1 Local System and Control Layer

Decentralized knowledge is among the key components of IoT. In decentralized knowledge, the prime target is to fabricate a medical gadget with canny control capabilities. This aides in handling operational data at the local level, in addition to the central server. These gadgets are usually enabled with sensors to measure operational parameters, converters to generate digital information sources, controllers to make real-time choices based on sources of info got from the converters, and system interfaces to share data with different machines/central servers. Examples of such gadgets are wearable screens, implants, and physician handheld diagnostic gadgets. Compatibility and the integration of advanced hardware are additional factors driving the utilization of IoT arrangements at the gadget level. Such gadgets are capable of acquiring qualitative, real-time biometric data from the patient's body and transmitting it under a tied down condition to a more significant level architecture. Encoders, actuators, and scrambling gadgets perform data transformations and pass it on to the following layer of the environment (for example communication conventions, for example, NFC and over-the-air programming) for analysis.

2.2.2 Devices Connectivity and Data Layer

The layer mainly centers around gathering data from the system gadget and putting away it in predefined data stores. The advancements at this layer are not one of a kind to any arrangement, (for example, patient checking). Verified medical data transfer advancements manage large data volumes and guarantee quality during the procedure. Systems administration firms, for example, Cisco and Oracle are very active in giving advanced innovations to end shoppers/framework integrators based on prerequisite.

2.2.3 Analytic Solutions

Layer Irrespective of the sorts of healthcare arrangements enabled, the central/remote server gathers data from different gadgets over the system and their key segments. The server with worked in algorithms analyzes real-time operational data to give bits of knowledge and ends. This data driven ingenuity assists with diagnostic ability, disease forecast, and actualizing preventive measures. The group and exhaustive evaluation of data from various sources, for example, implants and smart gadgets enables healthcare arrangements, for example, remote patient checking, intercessions, and incessant disease management.

3 Proposed Model

The entire process of the projected model is demonstrated in Fig. 1. In the beginning, the examination of the patients takes place using medical devices linked to IoT. When the MRI images of a person are acquired, pre-processing will takes place. Then, the features will be features from the pre-processed image. Simultaneously, class labels are provided to every image which is needed at the training procedure. Once the model is trained in an effective way, then the testing procedure will begin. Here, IGSAGA model is applied for filtering the features and optimal SVM model is applied for classification processes.

3.1 IGSAGA Based Feature Selection

3.1.1 Creation of Chaotic Series

Chaos is assumed as a prevalent occurrence in nonlinear model, which is a non-stochastic motion whereas it appears like a stochastic. The chaos series holds the

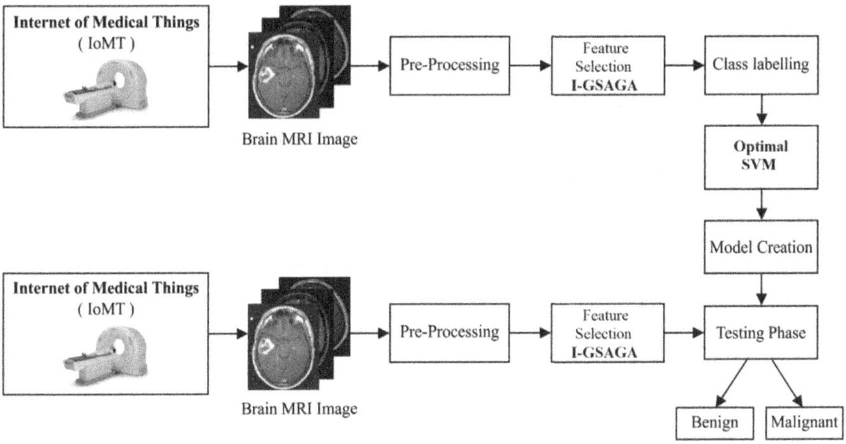

Fig. 1 Working principle of presented model

features of ergodicity and arbitrariness; hence it can discover every state with no redundancy in restricted area.

On comparing to the general arbitrary series, the chaotic series is similar to cover the global optimal region that offers are requirement for the model to attain the solution of maximum quality as well as precision. The IGSAGA model creates the chaotic series based on logistic equation and mapping takes place to the search space via carrier.

$$a_{i+1} = \mu a_i(1 - a_i), \quad 1 < i < n - 1 \tag{1}$$

where $a_1 \in (0, 1)$, $a_1 \neq 0.25$, $and\, a_1 \neq 0.75$, n indicates the entire population size, μ is the control variable of chaotic state. In case, when $\mu = 4$, the system is in absolute chaotic state.

3.1.2 Single Dimension Swimming

The general GSA model calculates the acceleration and the spirit of it is the consequential forces which act on agents from population as shown in Fig. 2.

Though it shows quicker convergence, it is easily trap to the local optimal. On the other hand, with respect to the GSA's full dimension movement mode, it is hard to attain maximum precision when the agents are placed near to the optimum value. Hence, the IGSAGA devises a movement of solitary dimension swimming, where swimming takes place using a step size s till no enhancement is attained. It is provided in Eq. (2), where d is single dimension index created in a random way.

$$s_i(t) = a_i^d(t) \tag{2}$$

$$x_i^d(t + 1) = x_i^d(t) + s_i(t) \tag{3}$$

3.1.3 Mutation Using t-Distribution

The t-distribution termed as student distribution, holds a degree of freedom as variable. When the degree of freedom reaches ∞, it is identical to the default Gaussian distribution, else when the degree of freedom is one, it is identical to the Cauchy distribution. The mutation based on t-distribution is employed on the initial k agents with largest weight in every round for additional diversification of the population and minimization of earlier convergence. It is represented as

$$v(t) = t_{max} - t + 1 \tag{4}$$

$$x_i(t + 1) = x_i(t)(1 + td(v)) \tag{5}$$

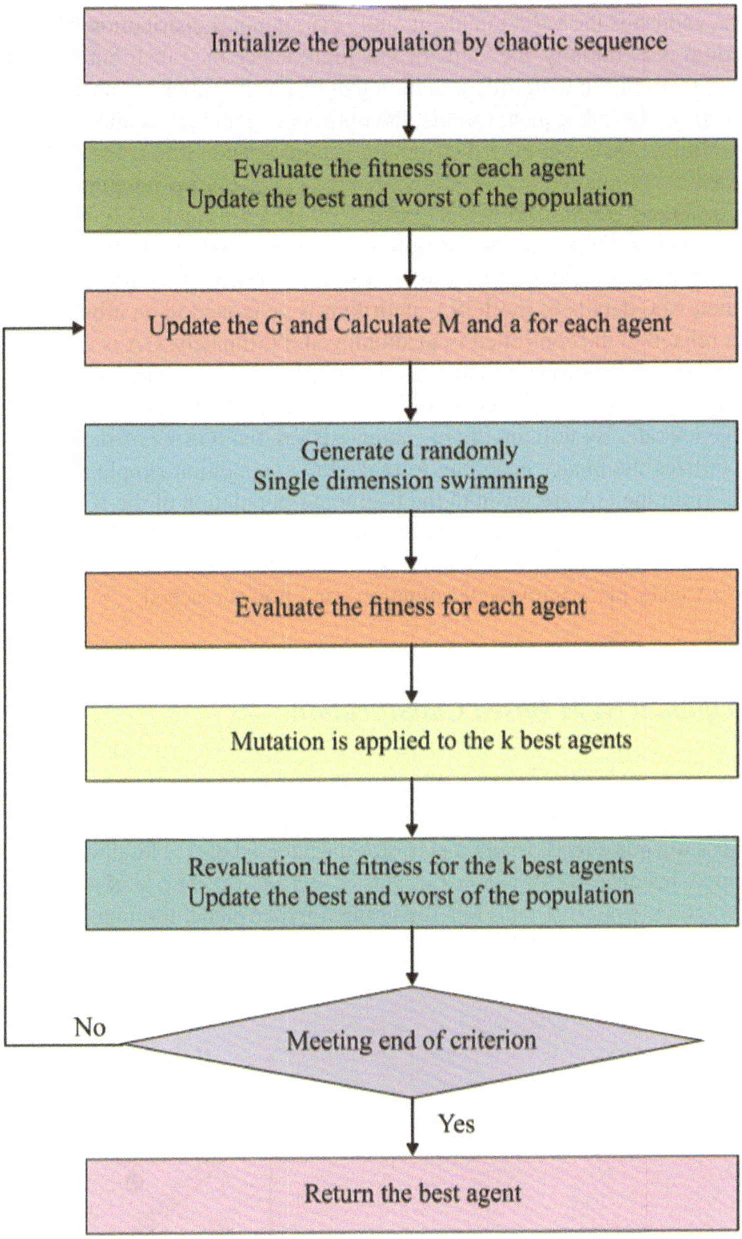

Fig. 2 IGSAGA based feature selection

where x_i indicates the variation agent, and $td(\upsilon)$ denotes distribution whose degree of freedom is υ. Initially, the value of v is high; hence the t distribution is equal to Gaussian distribution with little mutation that is advantageous to enhance the precision. Next, in the subsequent rounds, the optimum agents are attained the optimum regions. In that instance, the t distribution is corresponding to Cauchy distribution with maximum mutation that is advantageous to local optima jumping.

GA is based on the biological process of selection and evolution applied to generate meaningful solutions for optimizing and searching problems in case of absence of heuristic details of the issue to be controlled. It is considered that the initial heuristic details of the IGSA algorithm is generated in an arbitrary way and no data related to the controlled issue employed. Fortunately, GA is independent of any heuristic information during the evolution process and it could achieve possible solutions in various instances. At the same time, it is meaningless in the earlier convergence rate. By utilizing these characteristics, the IGSA-GA model is devised which utilizes the binary encoding to make the cooperation simple. The solutions attained from the GA are given to the beginning population of the IGSA model. In case of reaching to an efficient fitness value over GA, the individual is applied to replace the corresponding IGSA and GA for execution. Therefore, the execution of IGSA-GA takes place until the termination condition is reached.

3.2 *Optimal SVM Based Classification*

Once the features are filtered from the IGSAGA model, optimal SVM classifier gets executed to categorize the images to their respective classes. The SVM classifier model is shown in Fig. 3. Using a pair of n instance-labels, a classification problem is assumed, instance vector is $S = (x_i, y_i),\ (i = 1, 2, \ldots, n)\ x_i \in R$ and class labels are denoted as $y_i \in \{-1 +1\}$. For searching a hyper plane, training process of the classification model is applied that differentiates the negative as well as positive instances.

Fig. 3 SVM classifier

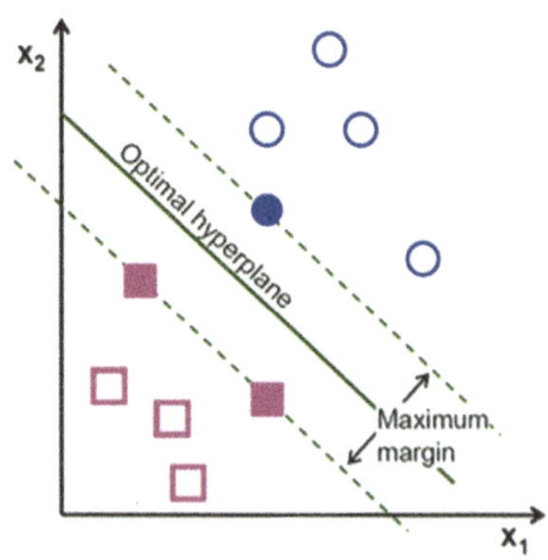

The optimization involved in the training process is provided in Eq. (6):

$$\emptyset(w, \varepsilon) = \frac{1}{2}\|w\|^2 + C\sum_{i=1}^{n}\varepsilon_i \tag{6}$$

Based on the limitations:

$$y_i[w \cdot x_i) + b] \geq 1 - \varepsilon_i, \quad i = 1, 2, \ldots, n \tag{7}$$

where the hyper plane normal vector is ω, slack variable is $\varepsilon_i \geq 0$ to measure the classifier error. For the error term $\sum_{i=1}^{n}\varepsilon_i$, C is a penalty factor and Ø is a function which performs mapping of the input space to a high-dimensional feature space. The whole process is an easily linearly separable issue in high-dimension situation; the conversion depends on the kernel function of SVM as defined below.

$$K(x_i, x_j) = \emptyset(x_i)^T \cdot \emptyset(x_j) \tag{8}$$

For dealing with the issue of linear separable ones, it is helpful to a maximum extent. The practical outline is extremely tricky for polynomial kernel function which may incur maximum computation time. It is applied for the recognition of the non-linear mapping for RBF kernel function and the required number of RBF kernel function parameters is low and it includes minimum complexity. When compared to other applied kernel functions, several experimental outcomes exhibited that RBF kernel function attained improved results. So, to classify the images, the RBF kernel function is applied to each dataset for identifying the optimum penalty factor C combination and kernel function variable δ.

4 Experimental Analysis

4.1 Dataset Used

The performance of the introduced IGSAGA-SVM model is tested using the benchmark freely accessible BRATS 2015 dataset comprising a collection of MRI brain images [18]. The experimentation takes place in MATLAB and the applied dataset comprises a set of 3 distinct brain MRI image sub-datasets such as Training, Challenge as well as Leader board. Here, the first two sub datasets are employed. The first one comprises a total of twenty High Grade Tumor (HGT), ten Low Grade Tumor (LGT) images along with the actual truth images from diverse professionals. The images present in the training set are applied to train the classifier model and the images in challenge subset are applied to test the model. A set of two class labels namely malignant and benign images are present in the dataset. Figures 4 and 5 shows the sample dataset images of benign and malignant classes.

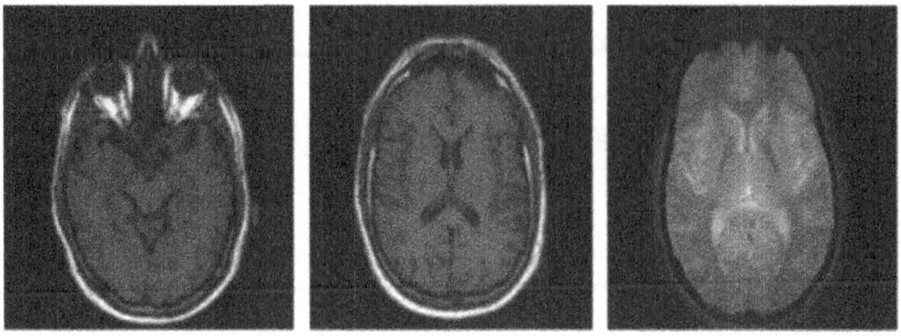

Fig. 4 Sample benign images

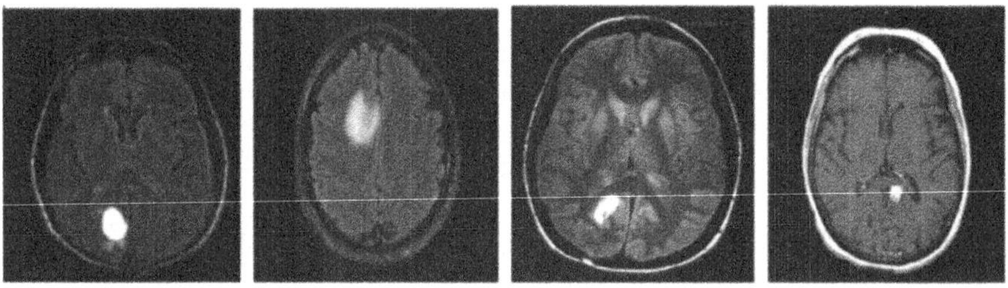

Fig. 5 Sample malignant images

4.2 Results Analysis

The IGSAGA-SVM approach performs proper recognition of the infected regions and also categorized it as benign and malignant. Besides, every identified region signifies the classifier model to categorize it as benign and malignant. Figure 6 shows the classification outcome of the presented IGSAGA-SVM model on the validation of the benign images. The first row denotes the actual images and the next row represents the classified benign images. Likewise, Fig. 7 demonstrates the classification of the IGSAGA-SVM model on malignant images. The first row denotes the actual images and the next row represents the classified malignant images. It is evident that the identified tumor areas and classification outcome with proper detection results.

Table 1 offered the comparison outcome of the projected IGSAGA-SVM approach with earlier models. Figures 8, 9 and 10 demonstrates the investigation of the projected IGSAGA-SVM approach with respect to sensitivity, specificity and accuracy in that order. Figure 8 displays the investigation of the projected IGSAGA-SVM approach with respect to sensitivity with the existing methods. Figure 8 states that the a highest value of 96.76% sensitivity is obtained by the projected model whereas the earlier model shown a maximum value of 96.20% only.

Figure 9 displays the investigation of the projected IGSAGA-SVM approach with respect to specificity with the existing methods. Figure 9 states that the highest value of 95.34% specificity is obtained by the projected model whereas the earlier model shown a maximum value of 95.10% only.

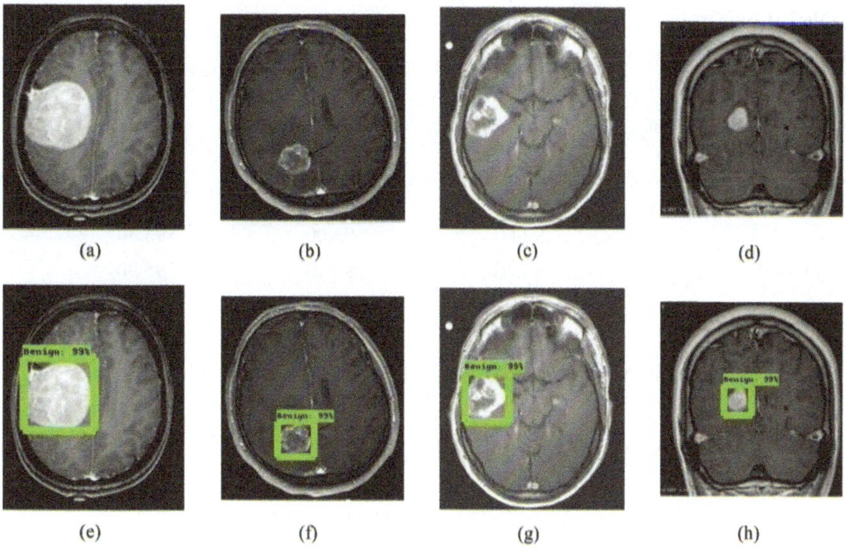

Fig. 6 Classification outcome on benign images

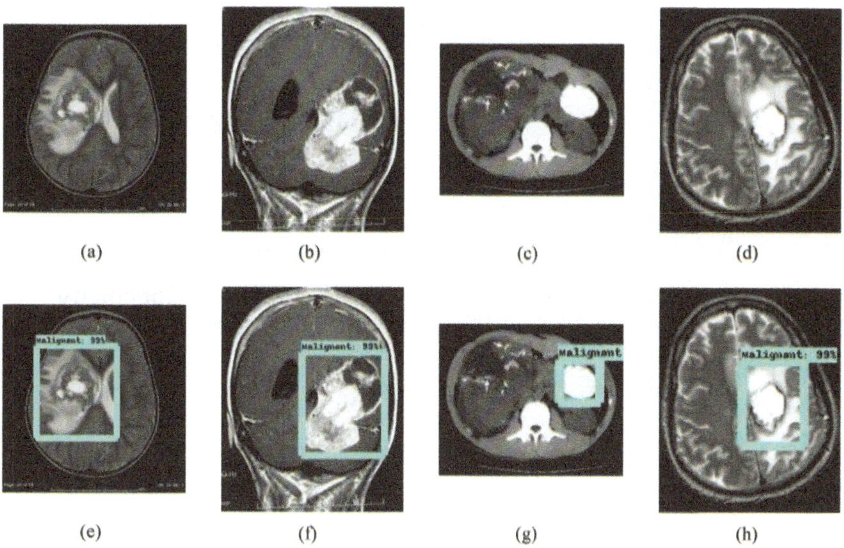

Fig. 7 Classification outcome on malignant images

Table 1 Comparisons of proposed with state of arts methods

Methods	Sensitivity	Specificity	Accuracy
Proposed	96.76	95.34	97.12
Selvapandian et al. [19]	96.20	95.10	96.40
Anitha et al. [20]	91.20	93.40	93.30
Pereira et al. [21]	94.20	94.40	94.60
Urban et al. [22]	92.60	93.00	93.30
Islam et al. [23]	94.30	95.10	95.90

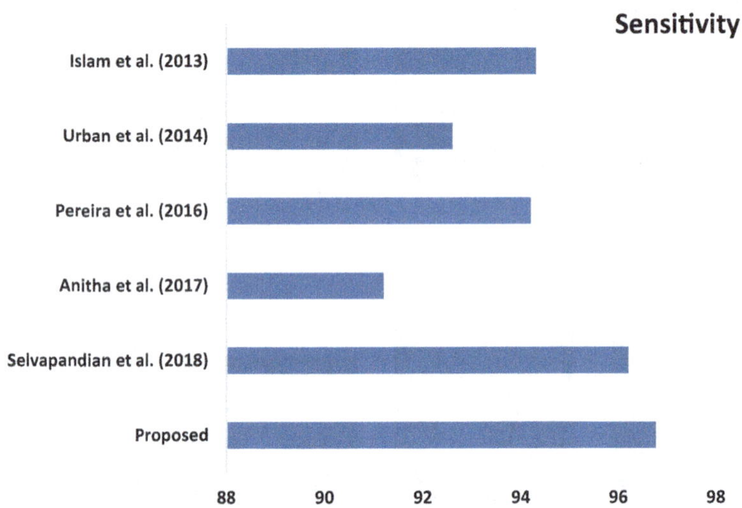

Fig. 8 Sensitivity analysis of distinct approaches

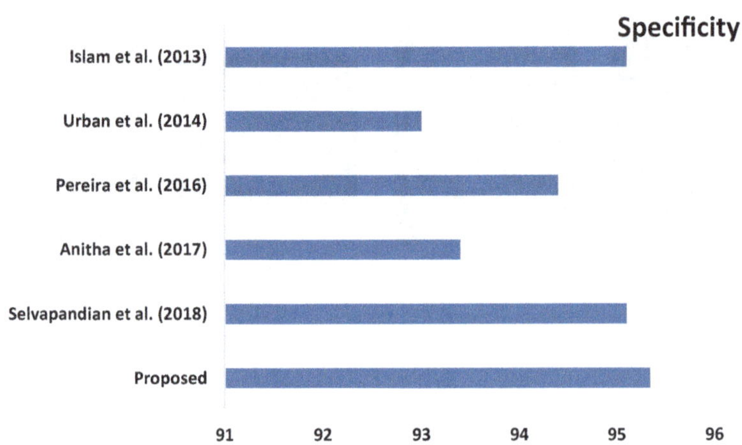

Fig. 9 Specificity analysis of distinct approaches

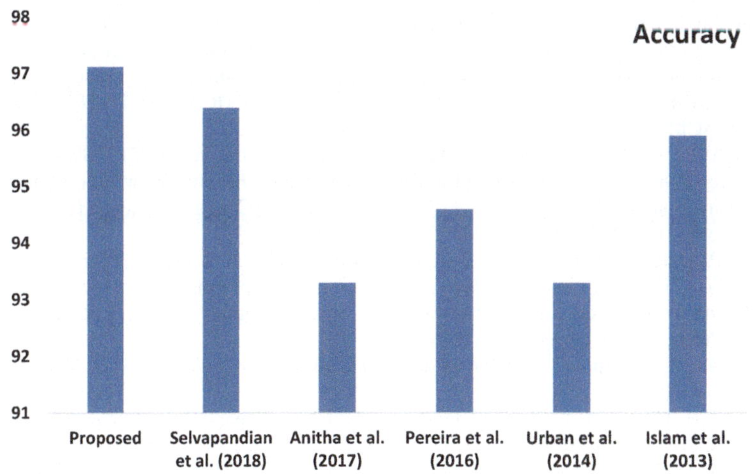

Fig. 10 Accuracy analysis of distinct approaches

Figure 10 displays the investigation of the projected IGSAGA-SVM approach with respect to accuracy with the existing methods. Figure 10 states that the highest value of 97.12% accuracy is obtained by the projected model whereas the earlier model shown a maximum value of 96.4% only.

5 Conclusion

A new Brain tumor diagnosis model is presented in this paper. Here, IGSAGA model is applied for filtering the features and optimal SVM model is applied for classification processes. The performance of the introduced IGSAGA-SVM model is tested using the benchmark freely accessible BRATS 2015 dataset. A set of two class labels namely malignant and benign images are present in the dataset. The IGSAGA-SVM approach performs proper recognition of the infected regions and also categorized it as benign and malignant. Besides, every identified region signifies the classifier model to categorize it as benign and malignant. The experimental outcome stated that the highest value of 97.12% accuracy is obtained by the projected model whereas the earlier model shown a maximum value of 96.4% only.

References

1. Elhoseny, M., Shankar, K., Uthayakumar, J.: Intelligent diagnostic prediction and classification system for chronic kidney disease. Sci. Rep. **9**(1), 9583 (2019)

2. Ganesan, M., Kumar, A.P., Krishnan, S.K., Lalitha, E., Manjula, B., Amudhavel, J.: A novel based algorithm for the prediction of abnormal heart rate using Bayesian algorithm in the wireless sensor network. In: Proceedings of the 2015 International Conference on Advanced Research in Computer Science Engineering & Technology (ICARCSET 2015), Mar 2015, p. 53. ACM
3. Sriram, R., Geetha, S., Madhusudanan, J., Iyappan, P., Venkatesan, V.P., Ganesan, M.: A study on context-aware computing framework in pervasive healthcare. In: Proceedings of the 2015 International Conference on Advanced Research in Computer Science Engineering & Technology (ICARCSET 2015), Mar 2015, p. 39. ACM
4. Paul, P., Dutta, N., Biswas, B.A., Das, M., Biswas, S., Khalid, Z., Saha, H.N.: An internet of things (IoT) based system to analyze real-time collapsing probability of structures. In: 2018 IEEE 9th Annual Information Technology, Electronics and Mobile Communication Conference (IEMCON), Nov 2018, pp. 1070–1075. IEEE
5. Ganesan, M., Sivakumar, N.: A survey on IoT related patterns. Int. J. Pure Appl. Math. **117**(19), 365–369 (2017)
6. Maitra, M., Chatterjee, A.: A Slantlet transform based intelligent system for magnetic resonance brain image classification. Biomed. Signal Process. Control **1**(4), 299–306 (2006)
7. Abbasi, S., Tajeripour, F.: Detection of brain tumor in 3D MRI images using local binary patterns and histogram orientation gradient. Neurocomputing **219**, 526–535 (2017)
8. El-Dahshan, E.S.A., Mohsen, H.M., Revett, K., Salem, A.B.M.: Computer-aided diagnosis of human brain tumor through MRI: a survey and a new algorithm. Expert Syst. Appl. **41**(11), 5526–5545 (2014)
9. Menze, B., Reyes, M., Jakab, A., Gerstner, E., Kirby, J., Kalpathy-Cramer, J., Farahani, K.: NCI-MICCAI challenge on multimodal brain tumor segmentation. In: Proceedings of NCI-MICCAI BRATS, Nagoya, Japan (2013)
10. Geremia, E., Zikic, D., Clatz, O., Menze, B.H., Glocker, B., Konukoglu, E., Shotton, J., Thomas, O.M., Price, S.J., Das, T., Jena, R.: Classification forests for semantic segmentation of brain lesions in multi-channel MRI. In: Decision Forests for Computer Vision and Medical Image Analysis, pp. 245–260. Springer, London (2013)
11. Shankar, K., Lakshmanaprabu, S.K., Khanna, A., Tanwar, S., Rodrigues, J.J.P.C., Roy, N.R.: Alzheimer detection using Group Grey Wolf Optimization based features with convolutional classifier. Comput. Electr. Eng. **77**, 230–243 (2019)
12. Lakshmanaprabu, S.K., Mohanty, S.N., Sheeba Rani, S., Krishnamoorthy, S., Uthayakumar, J., Shankar, K.: Online clinical decision support system using optimal deep neural networks. Appl. Soft Comput. J. **81**, 1–10 (2019)
13. Lakshmanaprabu, S.K., Mohanty, S.N., Shankar, K., Arunkumar, N., Ramireze, G.: Optimal deep learning model for classification of lung cancer on CT images. Future Gener. Comput. Syst. **92**, 374–382 (2019)
14. Shankar, K., Manickam, P., Devika, G., Ilayaraja, M.: Optimal feature selection for chronic kidney disease classification using deep learning classifier. In: 2018 IEEE International Conference on Computational Intelligence and Computing Research (ICCIC), Dec 2018, pp. 1–5. IEEE
15. Shankar, K., Ilayaraja, M., Deepalakshmi, P., Ramkumar, S., Sathesh Kumar, K.: Effective MKSVM classifier with LDA methods for brain tumor detection in MR images. Int. J. Recent Technol. Eng. (IJRTE) **8**(4S2), 987–992 (2019)
16. Elhoseny, M., Bian, G.-B., Lakshmanaprabu, S.K., Shankar, K., Singh, A.K., Wu, W.: Effective features to classify ovarian cancer data in internet of medical things. Comput. Netw. **159**, 147–156 (2019)
17. Lakshmanaprabu, S.K., Shankar, K., Ilayaraja, M., Nasir, A.W., Vijayakumar, V., Chilamkurti, N.: Random forest for big data classification in the internet of things using optimal features. Int. J. Mach. Learn. Cybern. **10**(10), 2609–2618 (2019)
18. BRATS 2015 MRI Brain Tumor Image Dataset Source: https://mrbrains18.isi.uu.nl/
19. Selvapandian, A., Manivannan, K.: Fusion based glioma brain tumor detection and segmentation using ANFIS classification. Comput. Meth. Prog. Bio. **166**, 33–38 (2018)

20. Anitha, R., Raja, D.S.S.: Segmentation of glioma tumors using convolutional neural networks. Int. J. Imaging Syst. Technol. **27**(4), 354–360 (2017)
21. Pereira, S., Pinto, A., Alves, V., Silva, C.A.: Brain tumor segmentation using convolutional neural networks in MRI images. IEEE Trans. Med. Imaging **35**(5), 1240–1251 (2016)
22. Urban, G., Bendszus, M., Hamprecht, F., Kleesiek, J.: Multi-modal brain tumor segmentation using deep convolutional neural networks. MICCAI BraTS (brain tumor segmentation) challenge. In: Proceedings, Winning Contribution, pp. 31–35 (2014)
23. Islam, A., Reza, S.M., Iftekharuddin, K.M.: Multifractal texture estimation for detection and segmentation of brain tumors. IEEE Trans. Biomed. Eng. **60**(11), 3204–3215 (2013)

An Effective Fuzzy Logic Based Clustering Scheme for Edge-Computing Based Internet of Medical Things Systems

V. Sellam, N. Kannan, and H. Anwer Basha

Abstract At present times, Edge-computing acts as a critical part in remote healthcare systems as hospitals deploy Internet of Medical Things (IoMT) for medicinal applications. A major limitation of edge computing based IoMT systems comprises the conservation of the energy in medicinal gadgets and increases the lifetime of the system. So, energy efficient transmission strategy is needed for healthcare system. Presently, different models have been presented for improvising the IoMT lifetime and clustering is treated as an effective model to achieve energy efficiency in medicinal applications. This paper presents an effective fuzzy logic based clustering technique for IoMT applications, called FC-IoMT technique. The presented FC-IoMT technique selects the Cluster Heads (CHs) based on five input parameters namely Energy, Distance, Delay, Capacity and Queue. By the use of FC-IoMT technique, the amount of energy consumption in the IoMT system can be significantly reduced. The proposed model has undergone extensive validation and the results ensured the superior results under several measures.

Keywords IoMT · Fuzzy logic · Clustering · Energy efficiency

V. Sellam (✉) · N. Kannan
Department of Computer Science and Engineering, SRM Institute of Science and Technology, Ramapuram, India
e-mail: sellamv@srmist.edu.in

N. Kannan
e-mail: kannann@srmist.edu.in

H. A. Basha
Department of Computer Science, RAAK Arts and Science College, Villupuram, Tamil Nadu, India
e-mail: hiiamanwerbasha@gmail.com

A. E. Hassanien et al. (eds.), *Cognitive Internet of Medical Things for Smart Healthcare*, Studies in Systems, Decision and Control 311,
https://doi.org/10.1007/978-3-030-55833-8_6

1 Introduction

In recent times, Internet of Things (IoT) has been analysed extensively and executed at diverse features of day-to-day life which is applied in modern cities to agriculture, and even healthcare to live a healthy lifestyle. IoT dependent structures provide a capable as well as planned tactics to monitor the patients at clinics; however, it can be used anywhere and anytime with automated systems. While IoT is used in medical sectors, then it is named as Internet of Medical Things (IoMT). An IoMT has progressive growth in medical domains by establishing the model at remote healthcare social benefits, and productive approaches are applied to observe and analyze the disease [1].

Because of the pervasive calculation of IoT, it is simple to manage and take care of various medical entities containing subjects, doctor's instructions, medications, as well as medical tools. With the combination of IoT and Machine Learning (ML) tends make an effective remote health monitoring, maximum number of programmed medical care is obtained with systematizing transfers are achieved in prior to humans. By the developed technique, recent healthcare models activate telehealth, telerehabilitation, telesurgery and telemedicine which provides remote intensive care and monitoring the patients at home and hospitals [2]. It will not best way to grow wearable tools to physiological size; however, it can be a requirement of current healthcare production for generating a whole network in that medical nodes connected for human body structures wireless body sensor network and then medical data is changed to medical cloud with implies through IoMT methods. A usual structure of IoMT based on healthcare is illustrated in Fig. 1. It has 3 important aspects of IoMT, namely, gateways, body sensor network and data cloud center. Currently, IoMT are determining numerous functions for supplying healthcare facilities to isolated stakeholders. The medical data procedure from medical nodes connected by human body is provided to apprehensive staff as well as comparatives for ensuring the patient's information from everywhere.

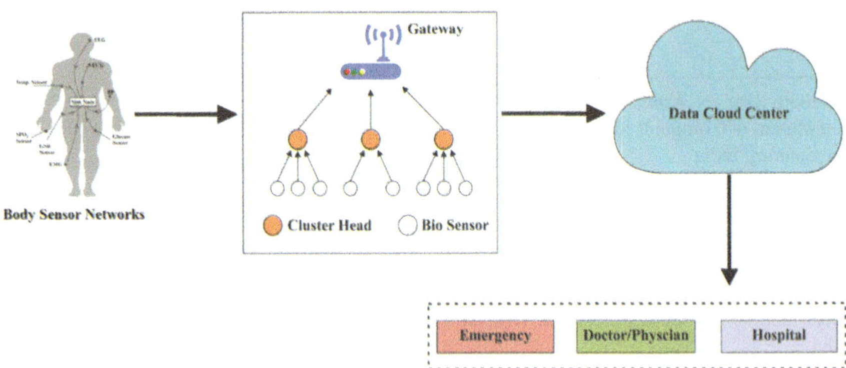

Fig. 1 General IoMT-based remote healthcare systems

The IoMT has the gateways named as central hub from medical tools and cloud data centre. Here, gateways are constrained with process of complete observation from smarter approach of health care, where the place of separate nodes is lined with the application of hospital. Then, the attained features are applied effectively by provisioning the gateways by adopting related intelligence, generating energy, as well as networking potential for creating modern gateways in telehealth monitoring. However, the applicable services are obtained using a gateway would be restricted while gateway is introduced in a independent manner. Additionally, the issue based on scalability and progressive development due to the productive elucidation distance management.

The situations needs additional layer to manage smart gateways to give intellectual facilities at the boundary of a model and enables to resolve the problems involved in medical sensors as well as CC layers. The obtained results are effective in communication named as edge computing [3]. It is allows the network to maintain frequent movement, reliability, and minimum delay as well as load-balancing, elements to develop remote health surveillance methods. Thus, the link of edge computing is combined with IoMT which produces value-relied healthcare at diverse medical sectors.

The edge computing is relied on IoMT functions; and the wearable nodes are included by valuable as well as restricted power sources. Energy consumption is a significant state of IoMT, it is referred to deploy energy-effective transmission modules to preserve the energy for every clinical devices that results in enhancing network lifetime. The clustering operation is majorly productive in sensor networks for applicable devices in IoMT [4]. It can roles an essential play to reduce require of several nodes involvement in the medical data broadcast to Base Station (BS) [5].

A medical tool is gathered with individuals sets of clustering mechanism that contains minimum central controller node named as Cluster head (CH), a further member in the set are cluster nodes (CN). The CN gathered medical data and broadcasts it to the CH for extra procedure. In cluster-adjusted schemes, the data broadcast series of BS is enhanced in case the selected CH contains minimum number of power. Besides, for IoMT functions, when the chosen of CHs is not correct, after further transmission in the cluster is needed that can require further energy utilization. Since, this issue reason difference energy process between medical nodes in IoMT.

The healthcare as well as smart homes is prevalent in the current scenario. So, many researchers are involved to grow an energy-efficient system to sensor networks. Duan et al. [6] projected the combined energy-efficient as well as consistent method depends on a game hypothetical technique for giving data protection in sensor networks. Initially, recognition of tools in IoT with danger policy depends on the manner. In addition, the overhead of the trust depends on manner is reduced with the game hypothetical technique. The outcomes make that they contain fitting effectiveness as well as safety to IoT based functions.

Kao et al. [7] present the 3 key involvements with presenting underwater channel method, its tasks as well as estimate subsequently. Sodhro and Sangaiah [8] present the combined IoT and their produce lifecycle management techniques for extending the battery duration and so power saving. In their modern methods are depends on the broadcast power manage as well as duty-cycle to controlling to optimize the battery charge utilization. However, it does not assume the wireless channels as well as energy optimize parameters and able cluster method at the network as well as physical layers. Deng et al. [9] presented the technique to balance delay in information broadcast as well as energy consumption to sensor networks. Also, they regard as the work-load organization systems to WSN.

Wang et al. [10] projected sensing layer depends on examining the energy dissipation in various sensors. Also, it creates a new structure according to the sensing, gateway, as well as control layers, however they duty-cycle as well as channel responsive energy stored in connection implement indicators as well as OSI layers. Kaur et al. [11] develops the new structure to the IoT functions mostly focuses at the 3 layers, for instance, sensing, and controlling, data process, as well as arrangement. However, it don't target on the cluster depends energy stored methods with assuming the several optimizing a signs on the network, MAC as well as Physical layers to the IoMT scheme. Heinzelman et al. [12] presented Low Energy Adaptive Clustering Hierarchy (LEACH) that are well identifying information broadcast and node chosen protocol to sensor networks. It executes a utilized arbitrary technique foe saving the energy of CH in the whole network. A CH chosen pursues the general rule it is several times the sensor node contains purposed as CH.

Mechta et al. [13] projected K-mean clustering technique to the sensor networks. A major issue by this technique is that, it realizes personalized centroid vector that outcomes in an inappropriate length of sets partition subsequent with losing connectivity. However, the benefits of these techniques get an overhead to develop cluster as well as variety of CH. This concern can be explained with implies of appropriate medium access control (MAC) method that establishes sleep mode of sensor nodes when they do not some information to broadcast. The MAC is extremely separated into 2 kind's contention-oriented as well as contention-free based manner. Thus, those protocols that perform contention are not suitable to massive methods. So, to solve these problems Time Division Multiple Access (TDMA) is established.

A TDMA enhances consistency as well as energy consumption ability of the sensor networks. The earlier clustering manners are depends on TDMA that does not consider the prospect of information worst in method. Thus, there makes a forecast overhead to perform clustering that achieves the resources method. It is developing of information failure estimate vaulting method to IoT schemes can solve aforementioned problem. Li et al. [14] projected interference-aware self-optimizing (IASO) to reduce existence of set-to station intervention. These techniques contain the ability of sensing multichannel together with the process for familiarize the control gain.

Qiu et al. [15] projected a Greedy Model with Small World (GMSW) to IoT functions. However, it requires extra growth for solving the problems. It is related such problems as a difficulty in resolving different issues, exclusive method, the restriction of particular channel estimate, and complexity in defining cognitive problems [16], etc. Moreover the CH collection in sensor networks is one of the most extraordinary techniques to storing energy between the medical nodes. Thus, it can be important for realizing the CH chosen process with implies of a fitting method through best convergence to develop energy effectiveness clustering in edge computing depends on IoMT.

A major limitation of edge computing comprises the conservation of the energy in medicinal gadgets and increases the lifetime of the system. So, energy efficient transmission strategy is needed for healthcare system. Presently, different models have been presented for improvising the IoMT lifetime and clustering is treated as an effective model to achieve energy efficiency in medicinal applications. This paper presents an effective fuzzy logic based clustering technique for IoMT applications, called FC-IoMT technique. The presented FC-IoMT technique selects the CHs based on five input parameters namely Energy, Distance, Delay, Capacity and Queue. By the use of FC-IoMT technique, the amount of energy consumption in the IoMT system can be significantly reduced. The proposed model has undergone extensive validation and the results ensured the superior results under several measures.

2 The Proposed FC-IoMT Model

In IoMT depends on remote healthcare, several medical tools are linked to all others. As the IoMT has numerous nodes, these networks are assumed to have "N" quantity of IoMT tools. The process of detecting and broadcasting the data to the respective IoMT tool that carries medical information to BS as depicted in Fig. 2. The necessary feature of the series that the tool is renovates the medical information on a tiny scale namely 2–5 m. In IoMT, a CH requires to be selected from different nodes in fog computing depends on IoMT.

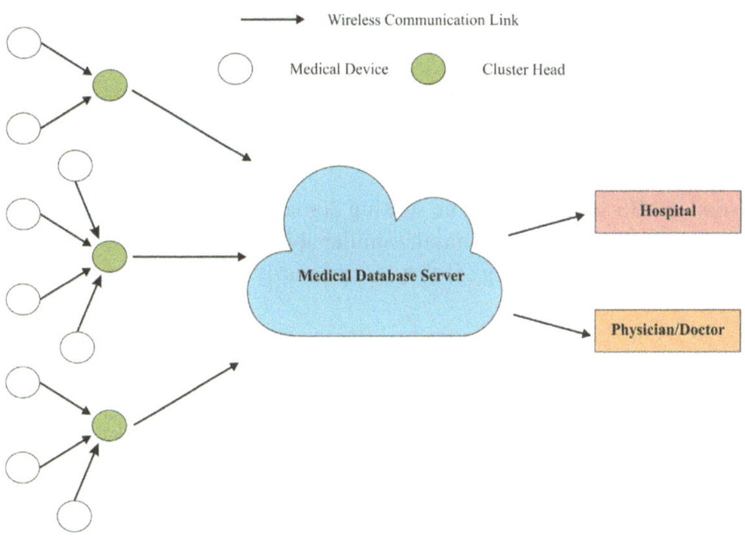

Fig. 2 Clustering framework for IoMT systems

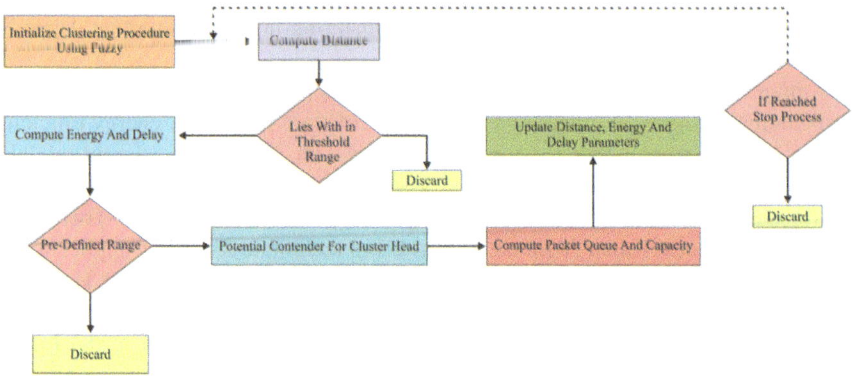

Fig. 3 Flowchart of proposed FC-IoMT model

In healthcare schemes, once medical tools are connected by the body, it can be not a simple role to reserve them. So, these studies meets that how to enhance the lifetime of the scheme, with the FC-IoMT aims to decrease energy utilization in the IoMT. The wearable healthcare schemes are in several methods similar to mobile ad hoc network (MANET) it is also set depends on mobility rather than sensor nodes depends on connection as well as movement. But, individuals cannot fit to the IoMT as further energy is needed. In the presented method gives the further extended network lifetime to escape several substitutions. The overall process involved in FC-IoMT model is shown in Fig. 3 [17].

2.1 Network Model

It is assigned with a sensor network which has n IoT gadgets arbitrarily used on a extensive application that observe the surroundings. It creates little utilization of sensor nodes and essential network approach:

- Sensors and BS situated from the sensing domain are kept stationary.
- A sensor is homogeneous and has the similar abilities. Every sensor node contains the similar quantity of energy if they are primarily used. All nodes are allocated a single identifier (ID).
- The sensors utilize power, manage to differ the quantity of broadcast power based on the distance to the required receiver.
- Links are symmetric. The node calculates the estimated distance to the other node according to the arriving signal strength, when the broadcasting power is identified.

2.2 Energy Consumption Model

It applies a simplify method to the transmission energy dissipation. Together the free space (d^2 power loss) as well as multi-path fading (d^4 power loss) channel methods are utilized, based on the distance among the transmitting as well as receiving. Energy spent to broadcast an l-bit packet on distance d is:

$$\begin{aligned} E_{Tx}(l,d) &= E_{Tx-elec}(l) + E_{Tx-amp}(l,d) \\ &= \begin{cases} lE_{elec} + l\varepsilon_{friss-amp}d^2, & d < d_0 \\ lE_{elec} + l\varepsilon_{two-ray-amp}d^4 & d > d_0 \end{cases} \end{aligned} \tag{1}$$

For receiving these messages, the radio uses energy:

$$E_{Rx}(l) = E_{Rx-elec}(l) = lE_{elec} \tag{2}$$

Between that $d_0 = \sqrt{\varepsilon_{friss-amp}/\varepsilon_{two-ray-amp}}$, the item E_{elec} signifies the energy utilization of radio dissipation, whiles $\varepsilon_{friss-amp}$ and $\varepsilon_{two-ray-amp}$ signifies the energy utilization to amplify radio that are utilized based on the broadcast distance.

It can be considered which the sensed data are extremely related, while the connection degree of sensed information is minimum, so the CH is always combined the information collected from its CN into unique length-fixed packet, also the function of data aggregation uses E_{DA}.

2.3 System Model and Assumptions

In the employed system, to execute FC-IoMT in IoMT; it can be considered with all medical tools are homogeneous and established randomly in the scheme. Besides, model assumes that corresponding quantity of energy are distributed to every connected medical tools, node in the intra-cluster contains arbitrary value through in pre-defined series.

The executed energy method states the usage of energy in data broadcast and reception methods by medical tools. Equation (3) signifies the formula to calculate power exhaustion of the data of length l bit to overall distance (d).

$$E_{T_s}(d) = \left(FS_T d^{\beta} + E_c\right)l \tag{3}$$

where E_c denotes the energy consumption of an digital circuit, FS signifies the receiver amplifier, and α indicates an model of path loss, $2 \leq \beta \leq 4$. Equation (4) reveals the power utilization with transmitter's node to obtain data packet.

$$E_R(d) = lXE_c \tag{4}$$

The entire energy utilization to all medial tool to transmit messages by distance "d" as depicted in Eq. (5).

$$E_{Total} = \{(FS)d^{\beta} + 2(E_c)\}l \quad (5)$$

2.4 Cluster Head (CH) Selection Process

From the network deployment phase, BS telecasts a beacon signal to each sensor node in a permanent energy level. Hence, every sensor is capable of computing an appropriate distance to BS that depends upon the attained signal strength. In line with several approaches, the process of newly projected clustering model has been classified into rounds as well as the features are assumed to be the Fuzzy Logic (FL). FL mimics the concept of human objective, that is minimum when compared with estimations that performs in a optimal fashion. Additionally, FL provides various exclusive features which makes an alternate method for managing the issues involved [8, 9]. In this approach, it has been defined with clustering process is performed in 5 stages.

1. Local Information Gathering. For every iteration, more number of tentative CH are selected in random manner for completing the last CH. The value of target CH is named as T, that is an pre-determined threshold. Then, a node which has not been selected as a tentative CH telecasts a HELLO_MSG with the given energy, and tentative node which obtains this message estimates an random distance to a transmitter and include them to neighbor nodes. Finally, 3 parameters were gathered as given below.
2. Fuzzification. Maps with desired input value to adjacent fuzzy sets and declares the truth value of MF for every fuzzy set. It is processed by providing values for all sets from Membership Functions (MF). The fuzzy sets as well as fuzzy MF are provided in Table 1.
3. Fuzzy Decision Blocks. The fuzzified rates are functioned by inference engine, that is comprised with rule base and diverse models to infer the rules. The rule base is defined a sequence of IF–THEN rules which combine the input fuzzy variables with resultant fuzzy variables under the application linguistic parameters, every

Table 1 Input variable and its fuzzy sets

Input variable	Fuzzy sets		
Energy	Low	Medium	High
Distance	Near	Moderate	Far
Delay	Low	Medium	High
Capacity	Low	Medium	High
Queue	Low	Medium	High

defined by a fuzzy set, and fuzzy implication operators AND, OR, and so on. Briefly, the 5 fuzzy input variables like energy, distance, delay, capacity and queue could be presented as x1, x2, x3, x4 and x5. Then, it is applied CHANCE as fuzzy output variable, that shows a tentative CH to be last CH and corresponding competing value. Then, denoted by y1 and y2.

$$\begin{aligned} Rule\,(i)\quad & IF\,x_1\,is\,A_1^i\,AND\,x_2\,is\,A_2^i\,AND\,x_3\,is\,A_3^i\,AND \\ & x_4\,is\,A_4^i\,AND\,x_5\,is\,A_5^i\,THEN\,y_1\,is\,B_1^i\,AND\,y_2\,is\,B_2^i \end{aligned} \tag{6}$$

where, i denotes ith rule in fuzzy rule base, A1 … A5 are fuzzy set of x1, x2, x3, x4 and x5. This rule base is comprised 243 rules. The rules are formed on the basis of criteria. Every rules in rule-base performed in concurrent fashion by fuzzy inference engine. The output is generated for all rules and results of rules aggregated finally. In case of simplicity, the model is applied with commonly applied fuzzy inference approach named as Mamdani Method.

4. Defuzzification. It is defined as the process of turning the result produced as well as collected by FL rules into a required output variable. It has several modules to processed, and it is applied with Centre of Area (COA) which is a basic and widely applied model as given below:

$$COA = \frac{\int \mu_A(x) \cdot x dx}{\int \mu_A(x) \cdot dx} \tag{7}$$

The corresponding fuzzy sets of resultant variables are given in Table 1. Every tentative CH computes the opportunity to be CH (CHANCE, CH) competition has been initiated. This CH telecasts a COMPETE_CH_MSG that is comprised with node ID and chance to become a CH inside a competence radius. The tentative CH identifies alternate tentative CH with a higher chance to become CH, it provides a competition by publishing a QUIT_COMPETE_MSG, and finally the tentative CH with maximum chance to become CH would be selected as CH. Else, it becomes a CH consequently. Once the CH is selected, every CH telecasts an broadcast message by a system. Normal sensors of a network combine with nearby cluster in LEACH and EEUC.

3 Experimental Validation

The performance of the proposed model has been validated using MATLAB tool and is tested using healthcare applications. The parameter settings involved in this simulation is provided in Table 2. A set of nodes undergo random deployment in the region of 300 × 300. Besides, medical nodes have to transmit data into PDA or smart phones lies in the range of 600 × 600.

Figure 4 shows the sustainability analysis of diverse models and the sustainability of the network can be estimated by determining the total number of alive nodes

Table 2 Parameter settings

S. no.	Parameters	Value
1	Initial energy	12 J
2	Network size	100 m × 100 m
3	Transmitting power	9 mW
4	Receiving power	3 mW
5	Data rate	190 kbps
6	Number of nodes	100

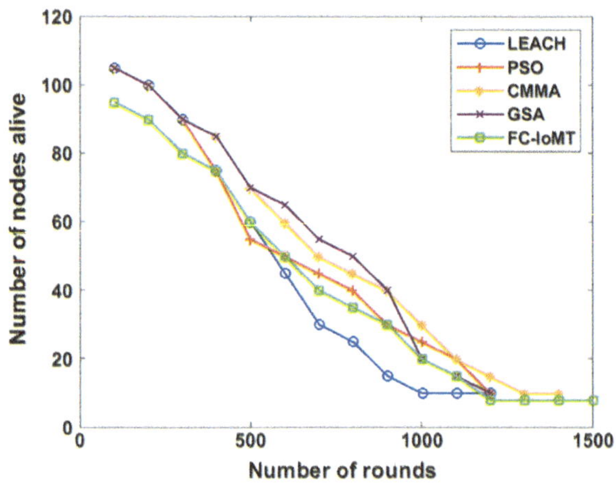

Fig. 4 Sustainability analysis of different models

existing till 1500 rounds. For example, the common idea is that the lifetime of the network enhances with the number of alive IOT devices existing in the network. Figure 4 clearly shown that the LEACH model has offered ineffective performance by attaining minimum network lifetime.

The allotment of medical nodes in every cluster by diverse models is shown in Fig. 5. It is shown that the LEACH model has offered ineffective results and requires more average number of nodes. It is also noted that the PSO algorithm has attained less number of nodes in a cluster over the LEACH model, but not effective than other CMMA and FC-IoMT models. At the same time, it is observed that the CMMA model has offered better outcome over the LEACH and PSO algorithms. However, the proposed FC-IoMT model has achieved superior outcome over the compared methods by requiring a minimal number of nodes under every CHs.

Figure 6 depicts the energy consumption analysis of diverse models under several number of nodes. Figure 6 showed that the LEACH model has achieved poor results by achieving maximum amount of energy utilization. In the same way, the CMMA model has resulted to moderate amount of energy utilization. However, the presented FC-IoMT model shows effective outcome over the compared methods in a significant way.

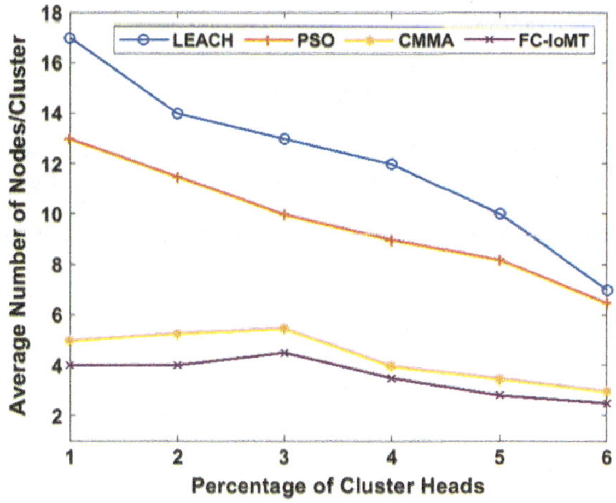

Fig. 5 Distribution of medical nodes in each cluster

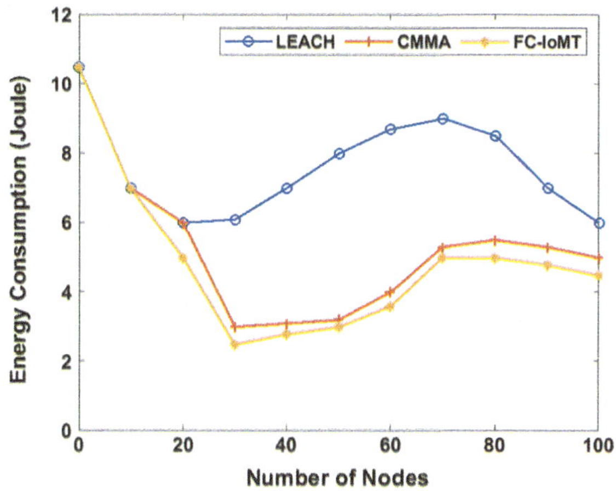

Fig. 6 Energy consumption analysis

4 Conclusion

This paper has presented an effective fuzzy logic based clustering technique for IoMT applications, called FC-IoMT technique. The presented FC-IoMT technique selects the CHs based on five input parameters namely Energy, Distance, Delay, Capacity and Queue. By the use of FC-IoMT technique, the amount of energy consumption in the IoMT system can be significantly reduced. The proposed model has undergone extensive validation and the results ensured the superior results under several measures. The proposed FC-IoMT model has achieved superior outcome over the compared methods by requiring a minimal number of nodes under every CHs.

References

1. Whitmore, A., Agarwal, A., Da Xu, L.: The internet of things—a survey of topics and trends. Inf. Syst. Front. **17**, 261–274 (2015)
2. Magsi, H., Sodhro, A.H., Chachar, F.A., Abro, S.A.K., Sodhro, G.H., Pirbhulal, S.: Evolution of 5G in internet of medical things. In: 2018 International Conference on Computing, Mathematics and Engineering Technologies (iCoMET), pp. 1–7 (2018)
3. Saad, M.: Fog computing and its role in the internet of things: concept, security and privacy issues. Fog **180** (2018)
4. Abbasi, A.A., Younis, M.: A survey on clustering algorithms for wireless sensor networks. Comput. Commun. **30**, 2826–2841 (2007)
5. Heinzelman, W.R., Chandrakasan, A., Balakrishnan, H.: Energy-efficient communication protocol for wireless microsensor networks. In: Proceedings of the 33rd Annual Hawaii International Conference on System Sciences, vol. 2, p. 10 (2000)
6. Duan, J., Gao, D., Yang, D., Foh, C.H., Chen, H.H.: An energy-aware trust derivation scheme with game theoretic approach in wireless sensor networks for IoT applications. IEEE Internet Things J. **1**, 58–69 (2014)
7. Kao, C.C., et al.: A comprehensive study on the internet of underwater things: applications, challenges, and channel models. Sensors **17**(7), 1–20 (2017)
8. Sodhro, A.H., Sangaiah, A.K.: Convergence of IoT and product lifecycle management in medical health care. Future Gener. Comput. Syst. (2018)
9. Deng, R.: Optimal workload allocation in fog-cloud computing towards balanced delay and power consumption. IEEE Internet Things J. 1–11 (2017)
10. Wang, K., et al.: Green industrial internet of things architecture: an energy efficient perspective. IEEE Commun. Mag. 48–54 (2016)
11. Kaur, N., et al.: An energy-efficient architecture for the internet of things (IoT). IEEE Syst. J. **11**(2), 796–805 (2015)
12. Heinzelman, W., Chandrakasan, A., Balakrishnan, H.: Energy-efficient communication protocol for wireless microsensor networks. In: Proceedings of the 33rd International Conference on System Sciences (HICSS 2000), p. 8020 (2000)
13. Mechta, D., Harous, S., Alem, I., Khebbab, D.: LEACH-CKM: low energy adaptive clustering hierarchy protocol with K-means and MTE. In: In IIT'14, Al Ain, pp. 99–103 (2014)
14. Li, Z., Chen, R., Liu, L., Min, G.: Dynamic resource discovery based on preference and movement pattern similarity for large-scale social internet of things. IEEE Internet Things J. **3**(4), 581–589 (2016)
15. Qiu, T., Luo, D., Xia, F., Deonauth, N., Si, W., Tolba, A.: A greedy model with small world for improving the robustness of heterogeneous internet of things. Comput. Netw. **101**, 127–143 (2016)
16. Duan, J., Gao, D., Yang, D., Foh, C.H., Chen, H.H.: An energy-aware trust derivation scheme with game theoretic approach in wireless sensor networks for IoT applications. IEEE Internet Things J.
17. Han, T., Zhang, L., Pirbhulal, S., Wu, W., de Albuquerque, V.H.C.: A novel cluster head selection technique for edge-computing based IoMT systems. Comput. Netw. **158**, 114–122 (2019)

Automated Internet of Medical Things (IoMT) Based Healthcare Monitoring System

Velmurugan Subbiah Parvathy, Sivakumar Pothiraj, and Jenyfal Sampson

Abstract The automated health care system in India is a necessity for health and the future. The population of the village is becoming more difficult due to the rapid growth of the healthcare population. Distant patients are far away from the doctor, but need constant monitoring and support. Resuscitation people can be at home or on the home. This article introduces an IoMT-based automated healthcare system for remote a patient that helps physicians and their connections. To predict the disease, the system conducts mechanical training using the CHAID algorithm (Chi-square automatic interaction detection Performs multi-level splits when computing classification trees) and generated multiple distributions during tree sorting to sort this data. If the outcome of the decision support system is the health of the patient, it will be notified by e-mail. This structure allows the doctor to intervene immediately to help patients with unusual health problems. The system sends alarms. By mail, if abnormal conditions are found in the patient's monitoring settings. The system successfully provides patients with intermediaries, nurses and relatives of patients in hospitals.

Keywords Healthcare monitoring · Classification trees · Wireless sensor network · Pulse oximeter sensor · Body temperature sensor · Heart rate sensor

---

V. S. Parvathy (✉) · S. Pothiraj · J. Sampson
Department of Electronics and Communication Engineering, Kalasalingam Academy of Research and Education, Krishnankoil, India
e-mail: s.p.velmurugan@klu.ac.in

S. Pothiraj
e-mail: siva@klu.ac.in

J. Sampson
e-mail: jenyfal.sampson@klu.ac.in

A. E. Hassanien et al. (eds.), *Cognitive Internet of Medical Things for Smart Healthcare*, Studies in Systems, Decision and Control 311,
https://doi.org/10.1007/978-3-030-55833-8_7

1 Introduction

Remote monitoring of well-being improves the well-being process and reduces the cost of services provided by the population. Analysis and recovery of the patient from the facial PSV minutes are the main features. Using innovations from the Internet of Things (IoT), the second statically validated face-to-face programming framework was proposed [1]. Internet of Things (IoT) is a promising invention for intelligent applications such as status monitoring and information development on the Internet. We are focusing on promoting recreational control based on the IoT system [2]. Relaxation is a unique condition for other people and different creatures. The problem of rest is the residual clinical problem of the individual or animal. We are making great strides in introducing the logic of well-being in front of the IoT door, so that IoT-dependent remote viewing facilities can be treated in the field of well-being, especially in remote and quite relevant regions [3]. The damaging effect of falls is of paramount importance in assessing, diagnosing and predicting health workers' health risks [4].

In order to monitor health and well-being, the IoT has achieved the highest standards of personalized health care, life support, customer positioning, collection and distribution of data at the discretion of the exercise [5]. IoT standards play an important role in improving personal well-being and well-being by expanding the availability and nature of medical care, as well as significantly reducing treatment costs and regular travel. To ensure cardiovascular performance, some scientists and planners have developed an ECG (Electrocardiogram) monitoring framework for patients with written access [6]. The Internet of Things is a system of interconnected devices, like a universal, compact and familiar device. IoT improves control synthesis with a number of applications. Correspondence between a large numbers of devices (any type of Internet association) can be provided via the Internet of Things [7]. One of the objectives of a viable, effective and safe remote monitoring system for patients is to summarize and record large flows of information without interruption, while protecting patient privacy. Information support is seen as a separate area, which includes the ability to store information and determine the level of accessibility for foreigners [8]. POINT-of-Care (PoC) is a patient test useful for predicting different conditions by evaluating simple biomarkers [1, 2, 9]. The main goal of PoC tests is to obtain rapid results from the start of treatment. From this point of view, silent observation can be improved at a lower cost [10].

Advances in computer-based innovations are growing day by day. The framework provides an emotional support network that examines the information received from the cardiovascular record, the patient's medical history, and continues to help predict heart problems at an earlier stage [11]. The leading cause of death is coronary heart disease. As a result, the picture focuses on expecting heart problems. In our methodology, a detailed structure for monitoring patients at home with the support of IoT with the help of fog has been created, which takes into account various occasions [12]. The height of the fog continues to determine the severity of the event, and the information currently selected sends a cloud for further investigation. The most important element of this tool is that the client can be anyone who needs to examine the patient, for example, a relative or specialist should not be with the patient without talking to the patient. The client can remotely check the patient's condition anywhere, the main obstacle being that the internet connection must be available to receive online updates about the patient [13]. Web-based cloud computing (IoT) is used to provide a planned pain assessment tool for patient remote control panels. The framework is designed for the ambulance clinic and for patients who are seriously considering another tool and use it to determine the strength of their pain as an option, as opposed to a face video [14]. Computer knowledge is distributed through a number of administrations to complete the apnea and perform treatment, including remote monitoring, continuous alerts, information retrieval, and data submission. The proposed framework allows clinical staff to agree on clinical decisions [15].

2 Literature Survey

Advances in Internet Security (IoT) security are still subject to significant changes and require a lot of updates. Due to device limitations, IT productivity is a major concern for IoT security. Understanding the observation is currently a hot topic of research. Rehman et al. [16] proposed a method to remotely assess a patient's ECG at or near a specialist. The main purpose of the examination is to differentiate the patient based on the EKG-based order. The ECG-ID Physionet (IDDB) database is used for ECG. The ratio of models to SSDs is used for validation (Sum of Squared Difference). The main characteristics of the characterization show the prospective fate of the created IoT security innovations.

As the Internet of Things (IoT) continues to improve, all the elements are interconnected, and this has been realized through the following innovative transformation. Some IoT applications include glossy shutdown, expert home, passionate city, harsh conditions, modern areas, rural areas and wellness experiments. One of these applications is well-tolerated screen well-being. Internet of Things makes clinical devices more efficient by improving the monitoring of the well-being of a permanent patient, where the sensor records tolerable information and reduces the likelihood of human error. The understanding of parameters on the Internet of Things is transmitted through the door through clinical devices, where they are stored and examined. The main difficulty in updating IoT for social security programs is the control of all patients in different areas. Therefore, the Internet or things in medicine offer a response to the successful observation of the cost of the patient, "to reduce the compromise between understanding the results and to cure the disease." Kumar and Rajasekaran [17] the Raspberry Pi card examined the patient's internal heat level, respiration rate, heart rate, and body development.

The current Internet Convention on Medical Services (IoT) is unsatisfactory in the creation of countries like Bangladesh due to the complexity, special support challenges and space costs. To solve this problem, Zilani et al. [18], presented a method based on financial R3HMS (Remote Reliable and Real-time Health Monitoring System), it is very easy to track and easy to perform key stop research In the clinical clinic of the area, during the hours of the staff of the Regional Clinical Medical Clinic, the patient may be: The range measures the patient's physiological indicators, such as ECG, airflow, and SPO2, which sends this extracted information to a remote server. The ESP8266 Wi-Fi chip is selected for long-distance correspondence and uses 8 pieces of ATmega328p MCU to collect and process information to reduce the pile in ESP8266. Using the Amazon Web Services (AWS) MQTT information convention to securely transfer information between devices. Authorities, specialists and various paramedics can regularly check this information to check the patient's condition.

Divakaran et al. [19], suggested that in today's clinical environment, the Internet of Things uses a wide range of sensors, so that experienced adults can access current clinical welfare administrations at any time. With the current welfare of the Board of Directors, the IoT innovation is reaching out to many clinicians and patients. Specialists can check and retrieve important patient information. The IoT is developing a smart letter strategy between the specialist and the patient. The company plans to design and demonstrate a hypothetical electronic remote health analysis framework that provides key clinical information and real-time videos that are distributed in the country. Urban centers are available in this area, which leads to better analysis. This patient is being treated.

The technical revolution to some changes in the field of "Internet of Things", all articles are assigned. The term IoT has been used in many areas, from the smart home to the smart city. In any case, the initial commitment of IoT to personal services is extraordinary. Kamble and Bhutad [20] presented a reflection on the well-being framework that refines the patient's physiological parameters such as temperature, pulse, and EKG, which can be connected to the Raspberry Pi card using these sensors. It sends information to its server. If you find unusual behavior, underlying signs, or side effects, the educator can be prepared by email or email, despite the success of the professionals. Improvement is important when developing a successful test plan at a distance. It then uses Shamir's cryptocurrency sharing calculator, which extracts information and keys, stores them in various fields, and uses them to obtain advanced passwords to ensure patient information security. Strengthening the support vector machine (SVM) projections used to supplement coronary artery disease. In this way, the framework offers first-class social security for all.

The tolerance review is currently part of a series of human services in medical clinics or at home. Rahman et al. [21] proposed a complex framework for patient testing that demonstrates patient well-being naturally using a variety of sensors. The information is then processed with precious data stored in the Raspberry Pi cloud IoT cloud. The frame actually removes the command mark from the ECG via the ECG sensor. Specialists (healthcare professionals/family members) can remotely monitor the patient's condition by constantly reviewing calm data and images. In the event that the condition does not become basic, a suggestion is sent to the specialist/guardian/comparator to instruct them and each meeting receives a video call to start this opportunity.

As suggested by Mathew and Abubeker [22], Worldwide, growing and unstoppable diseases are on the rise, from the medical clinic to the community of origin. It is based on continuous remote examination and examination of patients with Raspberry Pi 3. Raspberry Pi is an ARM11 frame chip. LINUX-based work frame for a visa the size of a single PC card. Python is a programming language used in Raspberry Pi, an open source programming language. This structure contains sensors whose organic parameters can be restored from the patient's body and remotely moved to a place that can be reached by any specialist. These settings are stored in the MySQL database and the purchased settings are processed in Pi and require a message. In previous strategies, the ZigBee and Bluetooth modules were used for transmission. However, they are limited to a short correspondence period. The chassis reduces human quality and costs.

3 Proposed Methodology

The two main goals of the system, which control the health of patients in the real world, are the main tasks of the system. The doctor or guardian can expect health parameters at anytime, anywhere in the system. Simultaneous signals are received when the health condition is abnormal. One of the problems of this system is the permission of data analysis to determine the database of the heart, the patient's medical history in time parameters. This is a planned disease system that informs the system when the outcome of the decision support system shows the best health status of the patient. The administration and the local control module deal with the rights of the system. This module is located at the change site of the patient. The data acquisition module accepted the input parameter values. The Wi-Fi module is used to listen to additional data on the web server. It will be appropriate for the patient's medical history. The data analysis module on the Doctors' Website is safe for managing CHAID classifier rules for course databases. All entries must relate to the decision support system. Three records, in particular the course database and the patient's medical history in general, will become the decision management system. To prevent the disease, this system conducts automatic courses based on these data to solve the CHAID classification algorithm of the decision tree. If the result of the decision is a decision on the patient's health, the decision is made by e-mail. With this option, your doctor can treat you immediately to help patients with abnormal health problems. Fig. 1 shows the overall proposed Automated IOMT based Healthcare Monitoring System.

3.1 Wireless Sensor Networks

A wireless sensor network is a remote system that consists of space-independent devices used by pointers to monitor physical or environmental conditions. WSN is

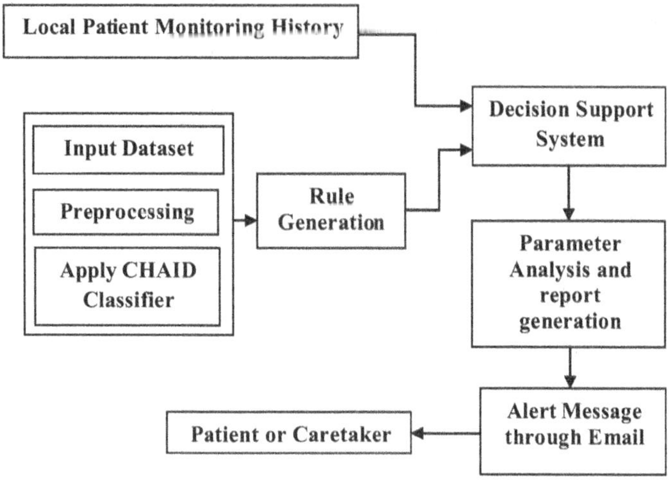

Fig. 1 Proposed automated IOMT based healthcare monitoring system

a system of innumerable sensor nodes, where each node is equipped with a pointer to distinguish physical light, such as light, heat, sound, etc.

3.1.1 Heart Beat Sensor

If the heart beat is indicated, it will emit a heart rate signal. If we integrate with the finger, the LED cools the LED with each heartbeat. The output of this computer can be connected directly to the microcontroller to evaluate the rhythm at any time Beats per Minute (BPM) conversion scale. It decomposes to adapt to light through all the blood.

3.1.2 Pulse Oximeter Sensor

Device description Pulse oximetry is a simple strategy for determining hemoglobin levels. The oximeter measures the number of pulses per stroke and is generally transmitted in bits per minute (Bpm).The program includes the use of a blood vessel oxidizer according to MCP6004, and the smart infrared optical sensor, TCRT1000, is used for photographic plethysmography (PPG). This technique is used to measure the mass of the heart, as the volume of blood is adjusted during the heartbeat.

3.1.3 Body Temperature Sensor

The internal heat level of a person is measured by the DS18B20 internal heat level indicator. To measure internal heat, the DS18B20 is placed under your hand, which is held under your fingers or worn on your forehead. DS18B20 internal heating level from −55 to 125 °C is a single-cable system for DS18B20 operation.

3.2 CHAID Algorithm

The CHAID selection tree calculation is used to verify the information. The rules are created from the UCI Heart database using the CHAID classifier. This calculation creates optional trees using Chi Square ideas to solve the perfect transport. CHAID is not at all like a container model and can create double-sided trees. This means that several units may have several branches. This is especially useful for studying large collections of information. The calculation of the agreed properties is converted to the standard using group calculations, as it works only with external or ordinary (direct) input data. The course uses agnostic techniques to test different key points "through objective results." It creates a tree diagram that deals with the subjects of the subject, most of whom expect an objective evaluation result. The CHAID provision is calculated as follows:

Binning Step The first step in calculating CHAID courses is to organize each permanent supply using the connection process, a_1, a_2, a_3, ..., a_{k-1}. For the progress of stable properties in A, which is characterized by an estimate of the difference given by 1, the model grouping the point of development, point $C(x)$ corresponds to the following (1).

$$C(x) = \begin{cases} 1 & x \leq a_1 \\ k+1 & a_k < x < a_{k+1},\ k = 1\ldots, k-2 \\ k & a_{k-1} < x \end{cases} \tag{1}$$

Merging Step Additional incentives should reduce the number of lessons in each property. For each power, the calculation of each pair of classes, which does not differ from the variable return, is combined by a chi-square model (with Pearson Chi-Square or Likelihood ratio). At this point, the calculation sets p (respect). Each failure of the two classifications is combined with one if the p is more significant than the $\alpha merg$; where, $\alpha merg$ the amperage is in the predetermined range [0, 1]. The value should be more significant than 0, not exactly, or equivalent to 1. However, if you ignore the fact that Ammerg's rating is set to 1, then this is the only consolidation level currently significant. This procedure is checked in turn to combine each pair of lessons that have the smallest difference. Existing property lessons can be combined without restrictions. However, for a typical brand (internal classes), only related classifications can be combined. The calculating CHAID calculates the balanced classification of grouped classifications using Bonferroni Compensation. Bonferroni coefficient B is the number of possible modes through which I can classify classifications to create r classifications. For $r = I$, $B = 1$. For $2 \leq r < I$, the simultaneous condition is used for 2 ≤ r < I.

$$B = \begin{cases} \binom{I-1}{r-1} \\ \sum_{v=0}^{r-1} (-1)^{v} \dfrac{(r-v)^{1}}{v!(r-v)!} \\ \binom{I-2}{r-2} + \binom{I-2}{r-2} \end{cases} \tag{2}$$

Splitting Step To find the best division, the calculation looks for the (best) division point with the lowest characteristic value. The retrieval procedure uses the added *P*-value to verify that each skill matches the performance variable. The lowest value tank is characterized by the highest level of ownership. If this value is not precise or equivalent to the default front $\alpha split$ separator, this force is selected as the current node separation work. After completing the separation procedure, the following subordinate centers are evaluated to determine whether they are equipped for separation or not.

Stopping Step Preparation for merge/separation continues many times until at least one last rule of each merged node is formed and no other units can be applied. This study uses the probability percentage to calculate the Chi Square test. The calculation creates a power table that contains the variables y of the classes of variable efficiency classes in the input features x as rows. Normal cell frequencies are evaluated differently without freedom of reception (i.e. no relation between x and y). Cellular frequencies are checked, and cellular frequencies are used to check the measurements of Chi-Squared statistic and the *p*-value.

$$G^2 = \sum_{j=1}^{j} \sum_{i=1}^{I} n_{ij} \ln\left(\frac{n_{ij}}{m_{ij}}\right) \tag{3}$$

When $n_{ij} = \sum_n f_n I(x_n = i \wedge y_n = j)$ cell repetition is observed frequency, normal cell $(x_n = i, y_n = j)$ $(x_n = i, y_n = j)$ repetition is normal and the p value is computed as follows, $p = pr(x_d^2 > G^2)$.

The advantages of this model are the following:

- Bases based on chi-square measurements;
- This is a measurable nonparametric model of free distribution;
- It can manage a wide range of information factors: presumably double, regular and sequential;

- Too fast to choose "large" trees with the ability to collect information makes it extremely popular in various applications;
- The model uses pre-heavy time. The concentrate can be isolated by measuring the criticality.

Despite these interesting points, this calculation requires a lot of information to get reliable results. Furthermore, there is a problem for regions to switch to several factors, which can distort the translation of the general meaning of the potential, clarifying the responses to objective factors. Furthermore, like other dynamic models, CHAID addresses the problem of instability.

4 Result and Discussion

The accuracy, precision, and recall are assessed using the attached equation.

Accuracy
We agree to the two measures commonly used to evaluate ML orders. The accuracy of P is measured based on confirmed information

$$P = \frac{TP}{(FP + TP)}$$

And the recall R is calculated as

$$R = \frac{TP}{(FN + TP)}$$

In the event that *TP* and *FP* do not coordinate an inappropriate number (i.e. a negative class of the positive class), *FN* is the right number in the wrong order test.

At the time of conceptualization, the basic information area of UCI Cleveland was selected from these three data sets in order to select the basic database. Deletes the database that divides the database into a test database that contains between 70 and 30% of the courses. At this stage, using the Bayesian naive agreement and the KNN calculations to get information, we obtained the results of the impact report shown in Fig. 3, the Cleveland UCI Center database with the calculated CHAID, which is more accurate than the various Calculations. Fig. 2 illustrates the proposed accuracy measures of the Cleveland Dataset.

Based on the accuracy of the calculations compared to the effects of the calculations shown in Table 1. Table 2 shows the quality of the CHAID calculations using the Cleveland UCI database.

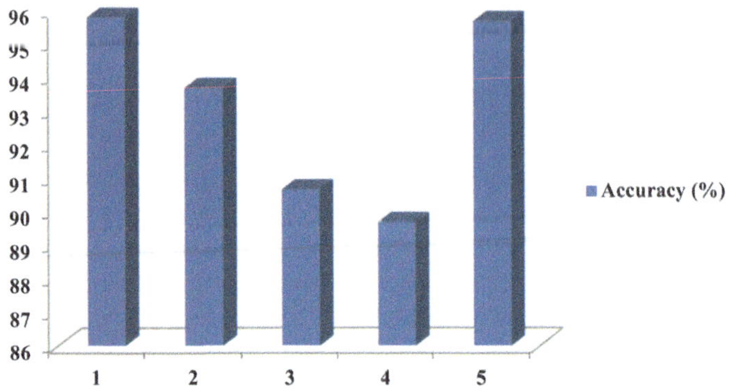

Fig. 2 Graphical representation of proposed accuracy measures

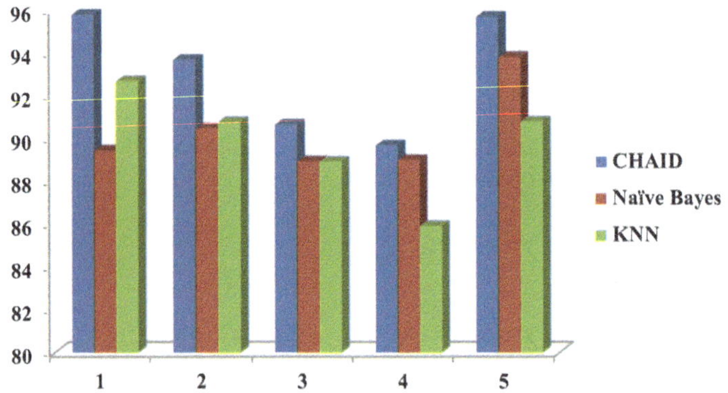

Fig. 3 Graphical representation of proposed and existing accuracy measures

Table 1 Proposed Cleveland dataset performance measures

Accuracy (%)	Precision	Recall
95.78	0.87	0.86
93.67	0.91	0.79
90.64	0.93	0.89
89.67	0.88	0.90
95.67	0.89	0.92

Table 2 Comparison of propose and existing accuracy measures

CHAID accuracy (%)	Naïve Bayes accuracy	KNN accuracy
95.78	89.43	92.67
93.67	90.45	90.76
90.64	88.90	88.89
89.67	88.96	85.89
95.67	93.78	90.78

5 Conclusion

Examining a calm patient is an incredible help for patients and social workers. In both cases, RPM is not yet available to anyone dependent on distance and remote access. Likewise, authorities should launch a special launch to end attempts to enter and use the RPM to encourage patients. The purpose proposed in this article depends on CHAID to determine patient IoMT, their health risks, and their severity. The structure recognizes typical and strange signals, arranges and recognizes irregular rhythms. It also advises professionals to provide better mental and clinical care at the right time. The normal overall accuracy of the proposed frame is 95.78% using the Cleveland UCI database and 0.93 for the typical prediction of the ECG sensor. In contrast to the work done [22], the picture has improved the precision of organizing typical and abnormal impulses. Further work focuses on the prognosis and transmission of the disease (from IoMT devices to cloud computing).

References

1. Bharat Kumar, G.J.: Internet of things (IoT) and cloud computing based persistent vegetative state patient monitoring system: a remote assessment and management. In: 2018 International Conference on Computational Techniques, Electronics and Mechanical Systems (CTEMS), Belgaum, India, pp. 301–305 (2018). https://doi.org/10.1109/CTEMS.2018.8769175
2. Lin, C., et al.: IoT-based wireless polysomnography intelligent system for sleep monitoring. IEEE Access **6**, 405–414 (2018). https://doi.org/10.1109/ACCESS.2017.2765702
3. Pathinarupothi, R.K., Durga, P., Rangan, E.S.: IoT-based smart edge for global health: remote monitoring with severity detection and alerts transmission. IEEE Internet Things J. **6**(2), 2449–2462 (2019). https://doi.org/10.1109/JIOT.2018.2870068
4. Saadeh, W., Butt, S.A., Altaf, M.A.B.: A patient-specific single sensor IoT-based wearable fall prediction and detection system. IEEE Trans. Neural Syst. Rehabil. Eng. **27**(5), 995–1003 (2019). https://doi.org/10.1109/TNSRE.2019.2911602
5. Satija, U., Ramkumar, B., Sabarimalai Manikandan, M.: Real-time signal quality-aware ECG telemetry system for IoT-based health care monitoring. IEEE Internet Things J. **4**(3), 815–823 (2017). https://doi.org/10.1109/JIOT.2017.2670022
6. Singh, P., Jasuja, A.: IoT based low-cost distant patient ECG monitoring system. In: 2017 International Conference on Computing, Communication and Automation (ICCCA), Greater Noida, pp. 1330–1334 (2017). https://doi.org/10.1109/CCAA.2017.8230003
7. Uddin, M.S., Alam, J.B., Banu, S.: Real time patient monitoring system based on internet of things. In: 2017 4th International Conference on Advances in Electrical Engineering (ICAEE), Dhaka, pp. 516–521 (2017). https://doi.org/10.1109/ICAEE.2017.8255410
8. Uddin, M.A., Stranieri, A., Gondal, I., Balasubramanian, V.: Continuous patient monitoring with a patient centric agent: a block architecture. IEEE Access **6**, 32700–32726 (2018). https://doi.org/10.1109/ACCESS.2018.2846779
9. Sood, S.K., Mahajan, I.: IoT-fog-based healthcare framework to identify and control hypertension attack. IEEE Internet Things J. **6**(2), 1920–1927 (2019). https://doi.org/10.1109/JIOT.2018.2871630
10. Afsarimanesh, N., Alahi, M.E.E., Mukhopadhyay, S.C., Kruger, M.: Development of IoT-based impedometric biosensor for point-of-care monitoring of bone loss. IEEE J. Emerg. Sel. Top. Circuits Syst. **8**(2), 211–220 (2018). https://doi.org/10.1109/JETCAS.2018.2819204

11. Utekar, R.G., Umale, J.S.: Automated IoT based healthcare system for monitoring of remotely located patients. In: 2018 Fourth International Conference on Computing Communication Control and Automation (ICCUBEA), Pune, India, pp. 1–5 (2018). https://doi.org/10.1109/ICCUBEA.2018.8697871
12. Verma, P., Sood, S.K.: Fog assisted-IoT enabled patient health monitoring in smart homes. IEEE Internet Things J. **5**(3), 1789–1796 (2018). https://doi.org/10.1109/JIOT.2018.2803201
13. Warsi, G.G., Hans, K., Khatri, S.K.: IOT based remote patient health monitoring system. In: 2019 International Conference on Machine Learning, Big Data, Cloud and Parallel Computing (COMITCon), Faridabad, India, pp. 295–299 (2019). https://doi.org/10.1109/COMITCon.2019.8862248
14. Yang, G., et al.: IoT-based remote pain monitoring system: from device to cloud platform. IEEE J. Biomed. Health Inform. **22**(6), 1711–1719 (2018). https://doi.org/10.1109/JBHI.2017.2776351
15. Yacchirema, D.C., Sarabia-JáCome, D., Palau, C.E., Esteve, M.: A smart system for sleep monitoring by integrating IoT with big data analytics. IEEE Access **6**, 35988–36001 (2018). https://doi.org/10.1109/ACCESS.2018.2849822
16. Rehman, A., Saqib, N.A., Danial, S.M., Ahmed, S.H.: ECG based authentication for remote patient monitoring in IoT by wavelets and template matching. In: 2017 8th IEEE International Conference on Software Engineering and Service Science (ICSESS), Beijing, pp. 91–94 (2017). https://doi.org/10.1109/ICSESS.2017.8342871
17. Kumar, R., Rajasekaran, M.P.: An IoT based patient monitoring system using Raspberry Pi. In: 2016 International Conference on Computing Technologies and Intelligent Data Engineering (ICCTIDE'16), Kovilpatti, pp. 1–4 (2016). https://doi.org/10.1109/ICCTIDE.2016.7725378
18. Zilani, K.A., Yeasmin, R., Zubair, K.A., Sammir, M.R., Sabrin, S.: R3HMS, an IoT based approach for patient health monitoring. In: 2018 International Conference on Computer, Communication, Chemical, Material and Electronic Engineering (IC4ME2), Rajshahi, pp. 1–4 (2018). https://doi.org/10.1109/IC4ME2.2018.8465482
19. Divakaran, S., Manukonda, L., Sravya, N., Morais, M.M., Janani, P.: IOT clinic-internet based patient monitoring and diagnosis system. In: 2017 IEEE International Conference on Power, Control, Signals and Instrumentation Engineering (ICPCSI), Chennai, pp. 2858–2862 (2017). https://doi.org/10.1109/ICPCSI.2017.8392243
20. Kamble, A., Bhutad, S.: IOT based patient health monitoring system with nested cloud security. In: 2018 4th International Conference on Computing Communication and Automation (ICCCA), Greater Noida, India, pp. 1–5 (2018). https://doi.org/10.1109/CCAA.2018.8777691
21. Rahman, A., Rahman, T., Ghani, N.H., Hossain, S., Uddin, J.: IoT based patient monitoring system using ECG sensor. In: 2019 International Conference on Robotics, Electrical and Signal Processing Techniques (ICREST), Dhaka, Bangladesh, pp. 378–382 (2019). https://doi.org/10.1109/ICREST.2019.8644065
22. Mathew, N.A., Abubeker, K.M.: IoT based real time patient monitoring and analysis using Raspberry Pi 3. In: 2017 International Conference on Energy, Communication, Data Analytics and Soft Computing (ICECDS), Chennai, pp. 2638–2640 (2017). https://doi.org/10.1109/ICECDS.2017.8389932

Deep Belief Network Based Healthcare Monitoring System in IoMT

B. Raghavendrarao, C. Sivaprakash, M. G. Gireeshan, A. Shajahan, and S. Prasanth

Abstract The Internet of Things is a new technology that devices can use to communicate on the Internet. The proposed work on Intelligent Next Generation Vital Signs Monitoring (i-NXGeVita) describes how the Internet of Things can be used for machine learning in healthcare. By connecting the Internet of Medical Things (IoMT) to a patient, doctors and nurses can monitor them around the world in real time. IoMT equipment collects body temperature, heart rate, heart rate (signals), ECG (electrocardiogram) sensor, temperature sensor and other important statistics. They are stored in the cloud and synchronized with the local server. Mechanical analysis includes data analysis, health risk detection and real-time weight analysis using algorithms. Based on the Deep Belief Network (DBN) and IoMT, i-NXGeVita can detect normal and abnormal heart rates and classify various defects. The implementation system is very helpful for doctors to understand the health risks and medically treat the best people also in this proposed study achieves the better accuracy for the healthcare

B. Raghavendrarao (✉)
Department of Computer Science and Engineering, Sri Sairam College of Engineering, Anekal, Bangalore, India
e-mail: raghavendrarao.cse@sairamce.edu.in

C. Sivaprakash
Department of Electronics and Communication Engineering, Sri Sairam College of Engineering, Bangalore, India
e-mail: shivadesk@gmail.com

M. G. Gireeshan
Jaibharath Arts and Science College, Kerala, India
e-mail: Mggireeshan@gmail.com

A. Shajahan
Department of Electrical and Electronics Engineering, Noorul Islam Centre for Higher Education, Kumarakovil, Tamil Nadu, India
e-mail: Shajahan.musaliar@gmail.com

S. Prasanth
Department of Instrumentation and Control Engineering, National Institute of Technology, Trichy, India
e-mail: sprasanthengg@gmail.com

A. E. Hassanien et al. (eds.), *Cognitive Internet of Medical Things for Smart Healthcare*, Studies in Systems, Decision and Control 311,
https://doi.org/10.1007/978-3-030-55833-8_8

monitoring system will provide a better accuracy such as 95.67, 90.56 and 95.76 measures.

Keywords Internet of medical things · Intelligent next generation vital signs monitoring · Deep belief network · Electrocardiogram

1 Introduction

It is imperative to observing of wellbeing parameters for individuals with changeless or genuine sicknesses so as to treat them in an auspicious way. Be that as it may, if the information is sent to a master or legitimate gatekeeper, it will do an extraordinary job. What's more, data security is a significant issue for this kind of structure since it contains significant and significant clinical databases [1]. In spite of the fact that the inexorably normal clinical parameters are stunning and exact, they don't meet certain essential prerequisites for present and future applications. Basic development is currently looked for forceful and summed up clinical organizations which have insignificant effect on the ordinary existence of the patient [2]. It is imperative to recognize human exercises so as to comprehend and break down human activities, in which various exercises are performed simultaneously. Because of the numerous potential varieties of the class, the acknowledgment of human movement is upsetting, prompting incalculable positions, enlightenment, deterrents, targets and speed [3]. Clinical expectation in people has demonstrated that this breathing is a significant instrument for precisely surveying the fundamental parameters related with cardiovascular capacity, supplanting customary breathing screens [4]. The present wellbeing bulletin has grabbed the eye of researchers and other clinical social orders. Different tests have been acted along these lines and numerous different advances are in progress.

The presentation of WSN checking lessens the expense of human administrations (thriving, time, number of associates, and so on.), offers chances to patients and encourages their every day practice during enrollment. Data on physical issues is given to wellbeing laborers. On account of continuous remote checking of utilizations, the transmission extend which is initiated promptly by the center can lessen power utilization, information move limit, execution load for server execution and the executives costs. Long haul cardiovascular wellbeing screens require different reliance figurings to precisely distinguish changes in PQRST structures, confounding the qualities of PQRST morphology, including ECG signal heartbeat, time vacillations, and primary deformities of the PQRST morphology, power-line impedance (PLI) Equipment-related energy [5]. In the Health and Health Observatory, Ireland is probably the best model for gathering and disseminating individual clinical consideration. The IoT is an arrangement of physical gadgets with all the gear in the framework. Each brilliant piece of the IoT is a "thing" that can be associated with the Internet [6]. Because of IoT, which incorporates a wide assortment of associations, it has empowered researchers and architects to make boards of specialists [7]. In splendid clinical applications, IoT sensors/gadgets are introduced to patients in an assortment of ways. Understanding prosperity (information) is recorded from ECG, fetal screening, temperature or glucose levels. The security of this data is crucial to the patient's life. The IoT gathering and human administrations experts center around securing every sensor and gadget in IoT facilitates while safeguarding the respectability of their data [8].

As of late, numerous projects have been made through IoT advancements that require a steady and dynamic investigation of data. The dark image of controlling a patient's prosperity is presently another thought [9]. Server haze decrease lessens the requirement for information move and improves hierarchical efficiency by giving constant information to widespread clients who are moving however much as could be expected [10]. Additionally, the structure presents four different ways of working and sorts out the exchange of reactions to specific patients. Physiological signs and the security of the situation of the old are recommended to be an image of clever thriving, both natural and arranged. Developments in the Internet of Things (IoT) give a modified apparatus to evaluating torment for remote patient perception offices. The proposed structure in this article is planned for hospitalized and concentrated patients who utilize another instrument to decide the power of particular trouble, rather than face video [11]. The proposed structure has been actualized adequately, and its training in the field of OSA control and treatment has totally changed. IoT framework incorporates IoT gadgets (sensors and actuators), related interfaces, IoT shows, and IT cloud interfaces [12]. Contingent upon the control of Internet-based human services, the remote arrangement of physiological information by patients relies upon an assortment of hierarchical models.

The remainder of the paper is organized as follows. Section II reviews the related work. Section III describes the Proposed Methodology. In Section IV describes Result and Discussion. Section V describes the conclusions and future work and at last references are given.

2 Literature Survey

Butros Saykali et al. [13] recommended that Internet of Medical Things (IoMT) specifically, control of utilization could diminish the approved and monetary weight on medicinal services faculty. Be that as it may, the absence of correspondence around it, the need to incorporate numerous structures, and security limitations are a genuine weight on the candidate designers. This article presents our work on refreshing the IoMT stage to respond to these inquiries. To affirm the thought, we likewise present a virtual application and a perception application made in our database.

On the off chance that you don't have the foggiest idea when to quit eating or eating excessively, it can cause numerous medical issues. In iLog Rachakonda et al. [14], offered an advantageous structure for following the client and as of now offering an additional measure of food. iLog gives data on the perspective of an individual: bunches that are taken care of by practices, for example, ordinary or disagreeable. Industrious pressure, uncontrolled or uncontrolled eating routine, and weight are firmly connected, paying little mind to certain neurological changes. We offer a profound learning model for desert IT stages, through which things can be seen, arranged, and estimated, normally, from the client's tablet. This investigation analyzes three reexamined principles that, as indicated by iLog, should be possible. iLog has refreshed two diverse minor stages. The means incorporate a generally utilized PDA and a protected personal computer, which can unquestionably be a piece of a framework that permits you to feel the different things iLog wear with glasses. The iLog model indicated an all out precision of 98%, with an ordinary exactness of 85.8%.

One of the fundamental issues of the present crisis facility is the absence of a fruitful test technique for improving a few patients. At the Emergency Clinic, where a few patients are being tried simultaneously, the structure of various observations for every patient is ineffectual and less regular. Clinical staff needs an establishment or system to consistently check the patient's condition and revive the patient's prosperity. Along these lines: IOMT (Medical Things Internet)—Sharifudin et al. [15], it is proposed to make a system that shows various parameters for persistent prosperity consistently and gives data to tolerant assessment and research. The structure permits specialists to inspect patients for a rescue vehicle center or home. The edge incorporates a temperature sensor and a heartbeat pointer to quantify the two principle powers recognized. NodeMCU is made on the card before the sensors send the evaluated data to the devoted server and evacuate it for additional conversation. The Web GUI application is utilized to show lasting data and the condition of prosperity. Also, this wellbeing screen alarms experts/parental figures if the deliberate worth is bizarre.

The present improvement of the conventional clinical model for participatory data can be coordinated through the Internet of Things (IoT) model, which contains sensors in its unique state (characteristic, advantageous and combined) for the prosperity of the customer to enact the "Remote Control Guide". The RF ID (RFID) is as of now being worked on. Some portion of the physical layer of the IoT offers singular administrations for individuals in touchy circumstances, with delicate, self-sufficient pointers and single-use sensors. Amendola et al. [16], an enormous number of accessible options are shown at the application level. A few instances of RFID can gather and procedure multifaceted data about human conduct as indicated by the guidelines of electrical and wellbeing accreditation. At long last, open issues and new research standards are examined.

Amin et al. [17], proposed a foundation of mental prosperity that joins cloud development with the Internet of Things (IoT). This office utilizes brilliant sensors for correspondence and more profound choices from a keen urban point of view. Emotional and insightful creation continually underscores business as usual and advises about exact, quick and amazing human administrations. To evaluate the unwavering quality of the proposed framework, we present the down to earth ramifications of the EEG pathology requesting technique utilizing inside and out preparing. We are continually utilizing a portion of the pointers of coordinated solace, including the brilliant EEG sensor and breaking down multifunctional comfort data. Quiet EEG signals are transmitted from shrewd IoT gadgets to the cloud, where they are handled and sent to a psychological module. The structure decides the patient's condition by analyzing unmistakable highlights, for example, appearance, discourse, EEG, improvement, and imagery. What to do the psychological module gets ready for the ceaseless actuation of the movement program. At the point when information is moved to a profound learning module, EEG codes are alloted as over the top or conventional. Status check and Understanding the aftereffects of EEG treatment is given to wellbeing experts who survey the patient's condition and offer crisis care when the patient is in genuine condition. The proposed inside and out preparing model offers preferred exactness over complex structures.

In the course of the most recent couple of years, the Internet you tuned in to, the appropriate responses you got, the ascent of new rights. This example of progress is to a great extent controlled by Radio Frequency Identification (RFID), Wireless Sensor Network (WSN) and Resistance Control. Having this model, Catarinucci et al. [18], consider an individual called IoTaware, a modified perspective on biomedical gadgets in patient, staff and crisis centers and nursing homes. As a rule with IoT vision, we offer Smart Hospital System (SHS), which interfaces with Relationships, Governance, RFID, WSN and Mobile Phone, CoAP/6LoWPAN/REST System Framework. With the arrangement of the Micro Crossing Force, Hybrid Sensing Network (HSN) has 6LoWPAN center points that improve and adjust to stable ecological conditions and patient physiological parameters and keep up UHF RFID abilities. The touchy data is checkpoint, where the essential knowledge application gives access to close by and potential clients through the REST web organization. There is little proof of individual rights to run the SHS proposition, with different individual chances and formative zones, while other advanced others increase gigantic ground.

Dey et al. [19], presented the improvement of remote advancement, which grew quickly because of its arrangement and arrangement for wired applications, considering the benefits of remote marker organize (WSN) programs specifically. These tasks are partitioned into a few sections, including government disability, clinical, mechanical and family unit PC frameworks. This audit considers the ECG remote control arrangement of the house, which utilizes the Zigbee development. Such structures can be helpful in observing individuals in their cases and normally checking them with fitting clinical contemplations, as individuals live longer in their baggage. Wellbeing offices can continually search for various physiological signs, propose further arrangements and perceptions. The attributes and obstructions of these hues can influence the wearer's psyche when considering the mandatory signs. Continually screen the sheets, measure and check the electrical gadget of the heart, which is offered to the purchaser. ZigBee gadgets can offer lower power, insignificant exertion and a basic reaction to control the ECG signal at home. Be that as it may, these structures are routinely separated because of the brilliant game plans of home control frameworks and families. This audit shows the best of the test as far as how well it presents the primary thoughts and substance of remote ECG screens. Also, ECG flagging models and suitability conditions are utilized. He additionally concentrated further challenges and openings. The report shows that these system conditions for checking the government assistance of the mass market are tended to with every presentation contrasted with the national observing and control structure.

3 Proposed Methodology

The offer includes the development of a health monitoring system (i-NXGeVita) that uses Internet of Things and machine learning algorithms. The ECG sensor recorded the patient's heart rate (heart rate signals). These signals are stored in a cloud system synchronized with the local server. The connection between the patient's cloud and the cloud is confirmed by IoMT devices such as Arduino Uno and Raspberry Pi 3. The DBN algorithm works with this data to distinguish between normal and abnormal heartbeats, which classify and distinguish abnormal rhythms. In addition, i-NXGeVita informs doctors in real time about the best possible medical care and assistance. The primary goal of the patient system is to reduce the number of consultations, to provide real-time care, to reduce mortality, to avoid sudden heart attacks, to reduce transportation costs, and to make medical diagnoses. Similarly, the primary goal of physicians 'careers is to monitor and control patients' health. In the end, he informs her at the right time to take care of the patient.

3.1 ECG Sensor (Module AD8232)

The ECG sensor collects the patient's pulse (heart signals). These codes are stored inside the cloud, which syncs with the nearest server. The task of the ECG sensor is to check the patient's circulatory currents (and movements) by measuring normal and abnormal signals. A typical ECG signal is presented as an intermediate period of PQRS-T, at which point the P wave begins: for the QRS complex, it rises 2–3 points and ends at 0.06–0.12 s. The PR interval ranges from 0.12 to 0.20 s. The entire public relations staff is crucial. The QRS complex follows a temporary PR of 5 to 30 points, ranging from 0.06 to 0.10 s. The ST segment extends from the S wave to the T wave. A typical T wave has a mass of up to 0.5 MV. The QT interval typically ranges from 0.36 to 0.44 s. This work is based on data from several databases, including the Massachusetts Institute of Technology-Beth Israel Hospital (MIT-BIH) Arrhythmia Database, MIT-BIH Malignant Ventricular Arrhythmia Database, and Arrhythmia.

Specifications

1. Single Lead Heart Rate Monitor AD8232 is the cheapest card used to measure heart rate.
2. This electric movement can be written in the form of simple ECG or ECG reading and execution.

3. The ECG can be very powerful, and the AD8232 single-step heart rate monitor is presented as a working speaker to help you effectively move away from the intermediate periods of PR and QT.
4. AD8232 is a barrier to the formation of ECG markers and other assessments of biological potential.
5. Aims to suppress, strengthen and target small bioenergy indicators, taking into account the serious conditions caused by the development of the terminal or a remote location.
6. The AD8232 heart rate monitor interrupts the connection of nine systems, which are based on various contacts, wires or connectors.
7. Enter SDN, LO + , LO-, U, 3.3 V, GND. This screen shows the contacts you need to work with Arduino or another Progress Board.
8. This board includes RA (right hand), LA (left hand) and RL (right foot) contacts, which can be used to interfere with personal performance.
9. In addition, there is an LED that responds to the rhythm of the heartbeat.
10. To use the heart screen, you will need a biomedical touchpad and a sensor connector.

Applications

- Monitoring Welfare and Sports Signals
- Versatile ECG
- Remote health screen
- Play equipment
- Bioelectrical signal collection

3.2 *Raspberry Pi 3 Model B*

The health monitoring system contains certain sensors that are associated with the patient and transmit data about the treatment network. Raspberry Pi is utilized as an aggregator and data processor. The patient and the expert utilize a cell phone/PC as a checking gadget. As in Fig. 1, the sensor outline is utilized to gather information or evaluations from patients, and the appraisals are changed over into signals. These

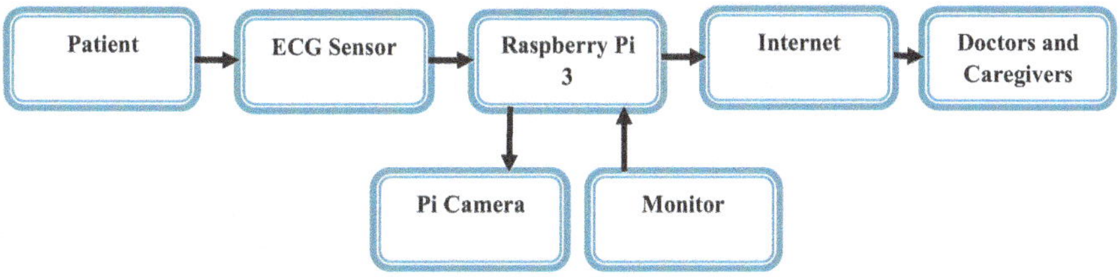

Fig. 1 Proposed healthcare monitoring system

characters are appropriate for dealing with the Raspberry Pi IoT module. Pi at that point shows the information on the screen and keeps it in the cloud. This information can be gotten to by an authority for his/her telephone/PC, and the information can be seen. In case of an emergency, the patient will at that point get a notice via mail to get the prescription. In the below section Fig. 1 describe our proposed healthcare monitoring system.

The equipment stage for the venture comprises of a sensor and Raspberry Pi 3 Model B, which is associated with the specialist by means of the Internet: advanced cell. This proposed thought will assist specialists with thinking about the patient's wellbeing and screen anyplace on the planet. These sensors gather the patient's clinical data, including the patient's pulse, circulatory strain, and pulse.

3.3 Preprocessing

We utilize the Pan-Tompkins estimation to address the reactions for the planning of crude ECG information (Offline and Online). There are two modules, for example, RSDetectorOffline and QRSDetectorOnline. RSDetectorOffline is utilized to recognize QRS structures in the data recorded by electrocardiography. QRSDetectorOnline is utilized to constantly distinguish ECG data in QGS structures. We concentrated on recognizing complex QRS issues, as this is the principle undertaking of the ECG test. QRS structures talk about the depolarization of the special, the left ventricle, and their ID is significant as a result of the flimsiness of the drive. Before partitioning the limit, the QRS complex is enlisted with four (4) strides in the Pan-Tompkins computation: Band-pass sifting, Differentiation, Squaring, Moving window combination, Thresholds modification. In Fig. 2 clearly explained the Working Process of Pan-Tompkins computation in the below section.

The handling steps of QRS complex location. P-wave is a managed work that began before the QRS complex. Along these lines, the P wave can be found relying upon the zone of the QRS complex. Container and Tompkins is one of the most famous QRS recognition mini-computers in practically all courses in the administration of natural clinical signs. Here is the computation chart. Figure 2 shows a diagram

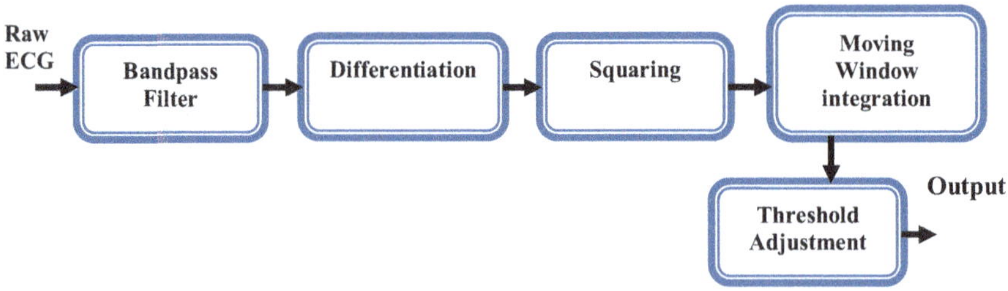

Fig. 2 Working process of Pan-Tompkins computation

of the primary computation time frames. The sign goes through the channel, subjection, square, and association stages, before characterizing the edges, QRS structures are seen. In the essential stage, the count is performed through a low entry and a high transient station to decrease torment with muscle constriction, troublesome impact of electric link, development of estimating gadget and T-wave impedance.

3.3.1 Band Pass Filtering

The channel intended to ascertain the QRS personality decreases the ECG signal clamor by considering the ordinary QRS unpredictability interim. This debilitates the skin because of muscle disarray, decay of the electrical cable, deviation from the benchmark and the impact of the T wave aggravation. Transmission power, which supports QRS reasonability, ranges from 5 to 35 Hz. The estimation in this count incorporates the high-pass and low-pass channels of the Butterworth IIR course.

3.3.2 Derivative Operator

The following degree of the executives is partition, a standard technique for deciding steep inclines, which normally perceives QRS structures from other ECG channels. The end procedure empties areas with a little redundancy of P and T waves and offers high recurrence because of the high inclines of the QRS complex.

3.3.3 Squaring

The action of the square gives a positive outcome and shows the tremendous logical inconsistencies that emerge from QRS structures. Because of the P and T waves, little complexities are wiped out. Rehashed indications of the high recurrence ascribed to the QRS complex have been additionally improved. This is a non-straight change that expects square sign examples.

3.3.4 Integration

The square wave passes through the slider window integrator. This addition summarizes the area under the snow wave at the appropriate time, enters the trial period, and includes the intrusion into the currently displayed window. Most of the half widths were chosen to be 27 to increase the life of the irregular QRS complex, but short enough to avoid both QRS complex and T wave. The MA (moving average) wave has a glow that does not fully correspond to the slope of the R wave. This ends with the completion of the corresponding contrast.

$$Y(nT) = 1/N[X(nT - (\mathrm{N} - 1)T] + \cdots + X(\mathrm{nT}) \tag{1}$$

where, N = 1 + 2M is the number of tests on the width of the skimming window. M is half width of the normal furniture wave. The onset of nervousness is frequent. The percentage of detection peaks is present, for example, sensitivity and efficiency the ability to perceive the calculation of Pan-Tompkins exceeds 99%. The payload is also low.

The low-pass filter is represented by the following equation:

$$y(n)\_ = 2y(n-1) - y(n-2) - x(n) - 2x(n-6) - x(n-12) \quad (2)$$

and the high-pass one is given by:

$$y(n) = y(n-1) - 1/32x(n) - x(n-16) - x(n-17) + 1/32x(n-32) \quad (3)$$

After sifting, the sign is divided to obtain data on the inclination of the QRS, as indicated in the attached formula.

$$y(n) = 1/8[2x(n) + x n - 1) - x(n-3) - 2x(n-4)] \quad (4)$$

Therefore, after the point, the sign is square, which means that all the information tricks become positive and higher frequencies occur.

$$y(n) = x^2(n) \quad (5)$$

3.4 Feature Extraction

After to deciding the multifaceted nature of QRS, we acquired key frequency, for example, the term QRS, while RR, greatest tallness R, most extreme development Q—the most noteworthy increment Q—the most noteworthy S recurrence, contingent upon the ECG. The fundamental data has been examined. The capacities recorded in the table above are utilized to set up our Deep Belief Network (DBN) to foresee whether the picture is normal of a heartbeat or not, and, thusly, to perceive the mood of abnormal rhythms. The common level, for instance, is as of now in the scope of 50–120 beats. Here we can recognize the normal and the surprising. In this examination, coronary supply route malady was depicted as common: Normal Beat (NB), Left Bundle Branch Block (LBBB), Sick Sinus Syndrome (SSS), Sleep Apnea Disorder (SAD), Fusion of Ventricular (FV), Premature Ventricular Contraction (PVC), Right Bundle Branch Block (RBBB), and Ventricular Fibrillation (VF). Normal Beat (NB), Left Bundle Branch Block (LBBB), Sick Sinus Syndrome (SSS), Sleep Apnea Disorder (SAD), Fusion of Ventricular (FV), Premature Ventricular

Contraction (PVC), Right Bundle Branch Block (RBBB), and Ventricular Fibrillation (VF).

Deep Belief Network

At Hinton's proposal, the deep belief network (DBN) got one of the most well-known inside and out learning algorithm [20, 21]. This is a short instructional exercise that can discover preferred settings over others. Fundamentally, the DBN test contains an uncontrolled arrangement plan that permits us to utilize Restricted Boltzmann machine (RBMs) as a restricted structure material, just as a vital recurrent layer for assessment. Restricted Boltzmann machine (RBM) is an ordinarily utilized stomatal sensory system that utilizations savvy preparing to manufacture the DBN. The RBM contains layer of double esteemed neurons and a layer of Boolean concealed neurons. There are reciprocal affiliations, even at various levels, yet there is no association between neurons at this level. The strategy for examining the situation of the layers to decide the likelihood contrast between the two levels relies upon their feasibility, as announced in (2). The likelihood of give up, in this way, can be expressed in (3).

$$\mathrm{E(v,h)} = -\sum_{\mathrm{i=1}}^{\mathrm{n_v}} \mathrm{a}_i v_i - \sum_{j=1}^{n_h} b_j h_j - \sum_{\mathrm{i=1}}^{\mathrm{n_v}} \sum_{\mathrm{j=1}}^{n_h} h_j W_{j,i} v_i \tag{6}$$

$$P(v,h) = \frac{\mathrm{e}^{-\mathrm{E(v,h)}}}{\sum_{\mathrm{v}} \sum_{\mathrm{h}} \mathrm{e}^{-\mathrm{E(v,h)}}} \tag{7}$$

where: vi is the number of neurons in the visible layer; hi is the number of Boolean hidden neurons within the hidden layer; $W_{i,j}$ is the weight matrices between the visible layer and hidden layer; and a_i and b_i are biases for the two layers. Next, the activation probability functions are presented in (4) and (5).

$$P(v_i = 1|h) = sig\,(\alpha_i + \sum_{j=1}^{n_h} W_{j,i}, h_j) \tag{8}$$

$$P(h_i = 1|v) = sig\,(b_i + \sum_{i=1}^{n_v} W_{j,i}, v_i) \tag{9}$$

where *sig*() shows the determined calculated sigmoid force. Subsequently, weight and pivoting boxes can be caused utilizing measures before them to can be made utilizing non-fake gear. Single concealed restricted Boltzmann machine (RBM) can't consolidate capacities with data. At long last, Boltzmann's DBN, a continuously developing transfer speed that is step by step growing in transmission capacity, is a redundant vital layer that can progressively expel huge abilities from the database.

Learning
Around there, the DBN is readied a few times as indicated by the properties got from the readied assortment (MIT-BIT data area) to perceive the Normal and Abnormal pulses thusly, to arrange coronary illness.

Evaluation
At the point when the model is executed precisely, it is assessed with the desire for immeasurable data (MIT-BIT data extend).

Prediction
In the event that the model is unsuitable after the assessment strategy, the readiness stage is rehashed to alter the planning data and to recreate a comparable system until the perfect outcome is accomplished. On the off chance that the introduction of the model is worthy, new data or current data can be normal.

i- NXGeVita screens the patient's condition utilizing Internet of Medical Things and Machine Learning (ML) Algorithms. An ECG sensor is applied to the patient to decide the degree of the heart, at that point it is sent to our closest server simultaneously through the cloud condition, Arduino Uno and Raspberry Pi 3 Model B.

The DNN calculation takes a shot at this information to decide tolerant well-being dangers and seriousness of appraisal among ordinary and strange pulses, sort and segregate between various anomalous rhythms. Also, the framework advises specialists progressively to give viable clinical consideration to the patient.

4 Result and Discussion

These assessments can be used to determine the overall accuracy and sensitivity of related conditions.

$$\text{Accuracy }(\%) = (\text{TP} + \text{TN}/\text{TP} + \text{TN} + \text{FP} + \text{FN}) * 100$$

$$\text{Sensitivity}(\%) = (\text{TP}/\text{TP} + \text{FP}) * 100$$

These conditions are also used to calculate the overall accuracy and sensitivity [22–24] of comparative coronary artery disease assessment.

In Table 1 and Fig. 3 result shows the accuracy measures for our proposed research in the given heart diseases such as NB, LLB, SSS, SAD and FV are trained and tested by the input dataset to attained the best accuracy such as 97.5, 98.4, 99.6, 98.3 and 99.7%.

In Table 2 and Fig. 4 the result shows the accuracy measures for our proposed research in the given Patient's Unique ID such as 3042020001, 3042020002 and 3042020003 and this proposed DBN attained the predicted accuracy such as 95.67, 90.56, and 95.76.

Table 1 Proposed research accuracy measures based on training and testing

Heart diseases	Training dataset	Testing dataset	Training accuracy (%)	Testing accuracy (%)
NB	16	16	100	97.5
LLB	23	20	100	98.4
SSS	19	19	100	99.6
SAD	27	25	100	98.3
FV	22	20	100	99.7

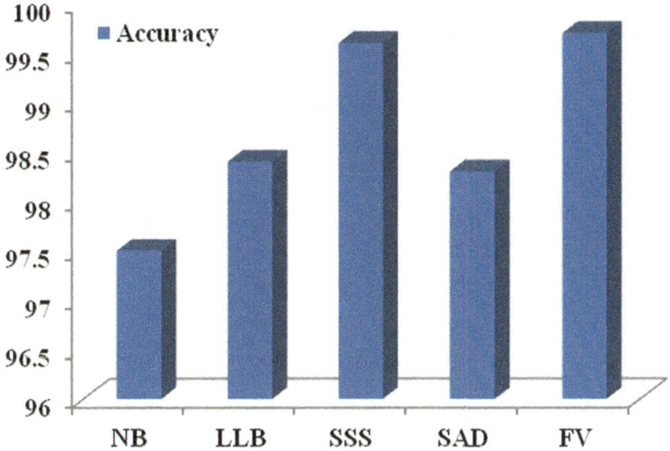

Fig. 3 Graphical representation of training and testing based accuracy measures

Table 2 DBN based accuracy prediction

Patient's unique ID	Accuracy of DBN prediction (%)
3042020001	95.67
3042020002	90.56
3042020003	95.76
Average	93.76

In Table 3 and Fig. 5 authors has discussed the proposed Deep Belief Network and existing Neural Network accuracy prediction measures. In this comparison will clearly show our proposed DBN will provide a better accuracy.

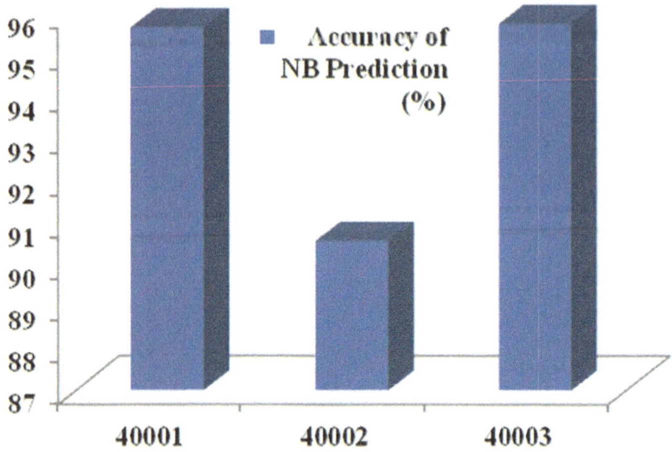

Fig. 4 Graphical representation of proposed DBN accuracy prediction

Table 3 Comparison of proposed and existing accuracy measures

Patient's unique ID	Accuracy of DBN prediction (%)	Accuracy of NN prediction (%)
3042020001	95.67	90.76
3042020002	90.56	87.76
3042020003	95.76	92.75

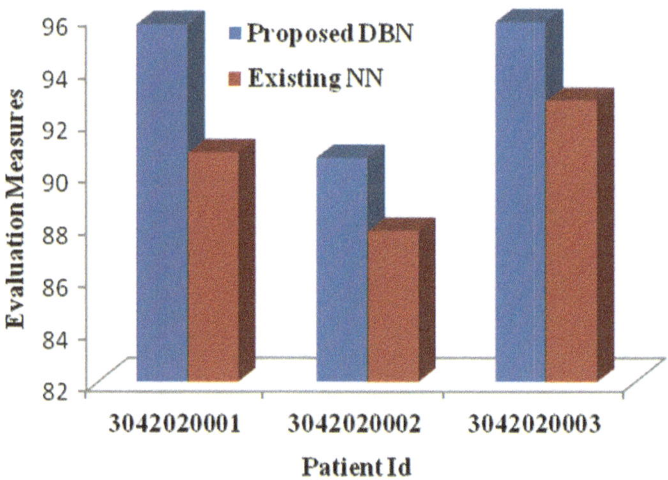

Fig. 5 Comparison of proposed and existing accuracy measures

5 Conclusions

The i-NXGeVita structure proposed in this article is a combination of a deep neural network (vital sign sensors) of the medical Internet of Things in combination with the main patient to collect evidence of dangers that are dangerous to his well-being, to the severity of the To determine ratings. It is not yet available for everyone. Professionals should also be inclined to involve patients and support the use of RBMs. After all, the main disadvantage of this improvement is the lack of light precision. It was suspected that the speed might be wrong for some people. The main idea of the proposed structure is to provide executives and clinical patients with an accurate and efficient well-being and to update the information cloud so that authorities and patients can use this data to implement their measures quickly and convincingly. The latest model is equipped with lighting that enables the specialist to observe his patient at any time and at any time. In addition, an ambulance or an ambulance signal can be created to inform the patient with complete medical information about the current state of the specialist. In future study is helpful for anticipating different illnesses with changing sensors prerequisites like ECG, EEG, and movement checking sensors and so on. The following of individual with GPS will be conceivable in emergency help condition.

References

1. Joshi. S., Joshi, S.: A sensor based secured health monitoring and alert technique using IoMT. In: 2019 2nd International Conference on Intelligent Communication and Computational Techniques (ICCT), pp. 152–156. Jaipur, India (2019). https://doi.org/10.1109/ICCT46177.2019.8969047.
2. López, G., Custodio, V., Moreno, J.I.: LOBIN: E-textile and wireless-sensor-network-based platform for healthcare monitoring in future hospital environments. IEEE Trans. Inf. Technol. Biomed. **14**(6), 1446–1458 (2010). https://doi.org/10.1109/TITB.2010.2058812
3. Gao, Y., et al.: Human action monitoring for healthcare based on deep learning. IEEE Access **6**, 52277–52285 (2018). https://doi.org/10.1109/ACCESS.2018.2869790
4. Li, C., Lubecke, V.M., Boric-Lubecke, O., Lin, J.: A review on recent advances in doppler radar sensors for noncontact healthcare monitoring. IEEE Trans. Microw. Theory Tech. **61**(5), 2046–2060 (2013). https://doi.org/10.1109/TMTT.2013.2256924
5. Satija, U., Ramkumar, B., Manikandan, M.S.: Robust cardiac event change detection method for long-term healthcare monitoring applications. Healthc. Technol. Lett. **3**(2), 116–123, 6 (2016). https://doi.org/10.1049/htl.2015.0062.
6. Satija, U., Ramkumar, B., Sabarimalai Manikandan, M.: Real-time signal quality-aware ECG telemetry system for IoT-based health care monitoring. IEEE Internet Things J. **4**(3), 815–823 (2017). https://doi.org/10.1109/JIOT.2017.2670022.
7. Sundaravadivel, P., Kesavan, K., Kesavan, L., Mohanty, S.P., Kougianos, E.: Smart-log: a deep-learning based automated nutrition monitoring system in the IoT. IEEE Trans. Consum. Electron. **64**(3), 390–398 (2018). https://doi.org/10.1109/TCE.2018.2867802
8. Tao, H., Bhuiyan, M.Z.A., Abdalla, A.N., Hassan, M.M., Zain, J.M., Hayajneh, T.: Secured data collection with hardware-based ciphers for IoT-based healthcare. IEEE Internet Things J. **6**(1), 410–420 (2019). https://doi.org/10.1109/JIOT.2018.2854714

9. Verma, P., Sood, S.K.: Fog assisted-IoT enabled patient health monitoring in smart homes. IEEE Internet Things J. **5**(3), 1789–1796 (2018). https://doi.org/10.1109/JIOT.2018.2803201
10. Wang, L., Hsiao, Y., Xie, X., Lee, S.: An outdoor intelligent healthcare monitoring device for the elderly. IEEE Trans. Consum. Electron. **62**(2), 128–135 (2016). https://doi.org/10.1109/TCE.2016.7514671
11. Yang, G., et al.: IoT-based remote pain monitoring system: from device to cloud platform. IEEE J. Biomed. Health Inform. **22**(6), 1711–1719 (2018). https://doi.org/10.1109/JBHI.2017.2776351
12. Yacchirema, D.C., Sarabia-JáCome, D., Palau, C.E., Esteve, M.: A smart system for sleep monitoring by integrating IoT with big data analytics. IEEE Access **6**, 35988–36001 (2018). https://doi.org/10.1109/ACCESS.2018.2849822
13. Boutros-Saikali, N., Saikali, K., Naoum, R.A.: An IoMT platform to simplify the development of healthcare monitoring applications. In: 2018 Third International Conference on Electrical and Biomedical Engineering, Clean Energy and Green Computing (EBECEGC), pp. 6–11. Beirut (2018) https://doi.org/10.1109/EBECEGC.2018.8357124.
14. Rachakonda, L., Mohanty, S.P., Kougianos, E.: iLog: an intelligent device for automatic food intake monitoring and stress detection in the IoMT. IEEE Trans. Consum. Electron. **66**(2), 115–124 (2020). https://doi.org/10.1109/TCE.2020.2976006
15. Sharipudin, A., Ismail, W.: Internet of medical things (IoMT) for patient healthcare monitoring system. In: 2019 IEEE 14th Malaysia International Conference on Communication (MICC), pp. 69–74. Selangor, Malaysia (2019). https://doi.org/10.1109/MICC48337.2019.9037498.
16. Amendola, S., Lodato, R., Manzari, S., Occhiuzzi, C., Marrocco, G.: RFID technology for IoT-based personal healthcare in smart spaces. IEEE Internet Things J. **1**(2), 144–152 (2014). https://doi.org/10.1109/JIOT.2014.2313981
17. Amin, S.U., Hossain, M.S., Muhammad, G., Alhussein, M., Rahman, M.A.: Cognitive smart healthcare for pathology detection and monitoring. IEEE Access **7**, 10745–10753 (2019). https://doi.org/10.1109/ACCESS.2019.2891390
18. Catarinucci, L., et al.: An IoT-aware architecture for smart healthcare systems. IEEE Internet Things J. **2**(6), 515–526 (2015). https://doi.org/10.1109/JIOT.2015.2417684
19. Dey, N., Ashour, A.S., Shi, F., Fong, S.J., Sherratt, R.S.: Developing residential wireless sensor networks for ECG healthcare monitoring. IEEE Trans. Consum. Electron. **63**(4), 442–449 (2017). https://doi.org/10.1109/TCE.2017.015063
20. Lakshmanaprabu, S.K., Mohanty, S.N., Krishnamoorthy, S., Uthayakumar, J., Shankar, K.: Online clinical decision support system using optimal deep neural networks. Appl. Soft Comput. **81**, 105487 (2019)
21. Lakshmanaprabu, S.K., Mohanty, S.N., Shankar, K., Arunkumar, N., Ramirez, G.: Optimal deep learning model for classification of lung cancer on CT images. Fut. Gener. Comput. Syst. **92**, 374–382 (2019)
22. Shankar, K., Lakshmanaprabu, S.K., Khanna, A., Tanwar, S., Rodrigues, J.J., Roy, N.R.: Alzheimer detection using Group Grey Wolf Optimization based features with convolutional classifier. Comput. Electr. Eng. **77**, 230–243 (2019)
23. Kathiresan, S., Sait, A.R.W., Gupta, D., Lakshmanaprabu, S.K., Khanna, A., Pandey, H.M.: Automated detection and classification of fundus diabetic retinopathy images using synergic deep learning model. Pattern Recogn. Lett. (2020)
24. Elhoseny, M., Bian, G.B., Lakshmanaprabu, S.K., Shankar, K., Singh, A.K., Wu, W.: Effective features to classify ovarian cancer data in internet of medical things. Comput. Netw. **159**, 147–156 (2019)

An IoMT Assisted Heart Disease Diagnostic System Using Machine Learning Techniques

K. Divya, Akash Sirohi, Sagar Pande, and Rahul Malik

Abstract Machine learning can be used across several spheres around the planet. The medical industry is not different. Health monitoring using wearable sensor enables us to go with Internet of Medical Things (IoMT). It enables the users to obtain the real time data i.e. live monitoring for manual prediction of user's health, using machine learning techniques. Data Generation is one of the most challenging problems which have been faced by many researchers. As the volume of obtained data is very large machine learning techniques need to be used. Machine Learning can predict the presence/absence of locomotor disorders and Heart diseases in our body. Such information, if predicted well ahead of time can provides essential insights to physicians who could subsequently schedule their treatment and diagnosis for their patients. In this paper, various machine learning algorithms have been implemented to predict the heart disease. 88.59% accuracy was obtained by using logistic regression with majority voting which is better than the existing techniques.

Keywords Heart diseases · Internet Of medical things (IoMT) · Machine learning · Majority voting

---

K. Divya · A. Sirohi · S. Pande (✉) · R. Malik
Department of Computer Science and Engineering, Lovely Professional University, Phagwara, Punjab, India
e-mail: sagarpande30@gmail.com

K. Divya
e-mail: ddivi786@gmail.com

A. Sirohi
e-mail: akashsirohi7@gmail.com

R. Malik
e-mail: malikvnit@gmail.com

A. E. Hassanien et al. (eds.), *Cognitive Internet of Medical Things for Smart Healthcare*, Studies in Systems, Decision and Control 311,
https://doi.org/10.1007/978-3-030-55833-8_9

1 Introduction

In order to track actual occurrences continuously, the Internet of Medical Things (IoMT) is an interconnection with different physical artifacts via sophisticated wireless network and sensors, the attached IoT devices interact. In the last decade, the Internet was used to transfer high-speed data. Intelligent fitness devices can be used by patients wanting to gather statistics, such as heart rhythms, blood pressure and glucose levels about their well-being. This information can also be tracked and sent to smart phones through sensors on wearable devices. IoT is used to get the real time data but if want to draw the manual data then IoT can be connected with the machine learning. The IoMT gadgets used for retrieving the health monitoring data is shown in Fig. 1.

Cardio vascular problems are the main reason for the heart diseases. These have a significant impact on the human health. The world's most common and mortal human disease is heart disease. In cardiac disease, the heart doesn't cooperate with other parts of the body. Sufficient amount of the blood is not being supplied to various parts of the body which results in irregular functioning of the body, hence causes heart failure. The symptoms of this conditions include lack of oxygen, body fatigue, feet get swollen and tiredness. The reasons for heart diseases are obesity, blood pressure, high cholesterol, smoking, eating unhealthy diet. The symptoms or features which are the reason for heart disease must be predicted in early stages so that one should cure it as soon as possible. The paper designs a prediction system of heart disease is done using classification techniques of machine learning. Before moving to classification lets overview various types and different ways the parts of body get affected by heart diseases [1]. The hierarchy of types of heart disease is shown in Fig. 2

Types of Heart Disease

- **Coronary artery disease**

Fig. 1 Raspberry Pi with the sensors

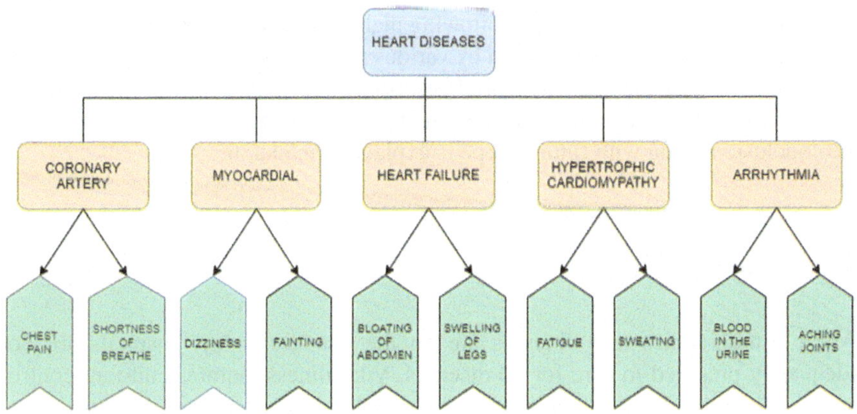

Fig. 2 Types of heart diseases

- Nutrients and oxygen stream through the circulation of the coronary arteries. The arteries can become ill or damaged by plaque deposits. This deposition of plaque affects the coronary arteries and reduces the oxygen and nutrients in the heart.
- **Myocardial infraction**
- Myocardial has various names like heart attack and coronary thrombosis. The heart muscles either get affected or completely killed by the abnormal blood flow. This particular heart disease caused by coagulation in a coronary artery that happens without person knowledge.
- **Cardiac Arrest**
- As the heart does not successfully pour the blood within the entire body which result in heart failure. The left and right part of the heart gets devastated. The heart becomes too rigid in case of high blood pressure or coronary artery disease.
- **Hypertrophic Cardiomyopathy**
- The thickening of left ventricle makes difficult for blood to flow out of the head. This disease is a genetic disorder. A person with hypertrophic heart disease has 50 percent chance that their offspring inherit this disorder.
- **Arrhythmia**

The problem arises because of improper working of hear-coordinating impulses. It makes the heart pound, whether it's too fast or slow. It is quite normal to have irregular heartbeats. The weakened heart or poor heart need to be got alter and handle smoothly.

Lot of research has been done in this area. Many researchers have used several machine algorithms with majority voting to obtained good results. But they faced the challenge of time consumption as they have implemented this technique by combining the algorithms. So a novel technique is proposed with five different machine learning algorithms which implements majority technique independently and provides the most promising result.

The chapter is organized in the following manner. Section 2 explains the working of different methodologies proposed by various researchers. Proposed Methodology is explained in Sect. 3 along with a brief description about dataset. Section 4 explains about classification techniques. Result Analysis is explained in Sect. 5. And finally the conclusion along with future scope is explained in Sect. 6.

2 Related Work

When demographics and illnesses are on the rise every day, hospitals are not adequately prepared to care for all diseases. Vital illness requires constant control on the condition of the patient's wellbeing, such as cholesterol rates, heart failure, etc. Using IoMT patients can be supervised at any time from remote location by a single physician. IoMT framework allows a patient to be remotely tracked. The live health data of the patient is captured and authenticated through the detectors and sent to the doctor. The guardian can track the health condition of patient and, in case of an emergency, can call the doctors directly. This preserved data of the patient and can allows the communication as well as diagnosis in emergency situations. Figure 3 depicts the implemented framework using IoMT.

As stated in [2] IoT has been combined with the machine learning to predict the heart disease as early as possible. Heart problems are not like simple cold or cough that somebody gets affected through air or other medium. In fact, heart problem prevails even from some childhood like blocking in heart,asthma problem etc. Detecting it at earlier stages will help the doctor to give the best treatment. For prediction various

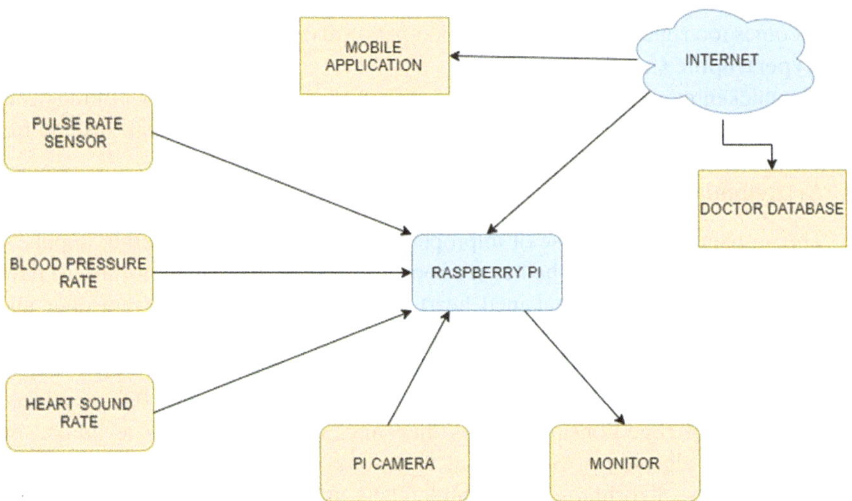

Fig. 3 Architecture of IoT framework

researchers worked on it, with different methodologies which have been discussed below in Table 1.

Shashikant et al. [3] develops a method which collected datasets from the analysis team of MituSkillogies. The aim of the system is to find the cardiac failure of a patient. It has three statistical models where 19 HRV inputs and two output groups. The terminologies like accuracy, precision, flexibility, specificities and AUC have been used to test these models. 93.61% accuracy is obtained by random forest algorithm.

Mezzatesta et al. [4] designed a predicting system for cardiovascular diseases in the patient. The author applied machine learning algorithms on two datasets that are Italian datasets and American datasets. Various approaches are used among all the applied methods, the best results given by Support vector machine with the accuracy 95.25% for Italian datasets and 92.15% for American datasets.

Beunza et al. [5] compares the classification or regression algorithms for forecasting clinical events. A Framingham heart study database is used to analyze the method. The methods like decision tree, random forest, support vector machine, neural network, and logistic regression,out of all the algorithms the author obtained the highest AUC value 0.75 by using the Support vector machine.

Beulah Christalin Latha et al. [6] developed a predicting system for analyzing the risk factor for heart diseases using ensemble classification techniques. The author not only deals with the accuracy but with the dataset too by implementing in order to predict in earlier stages. The author applied bagging and boosting which are ensemble methods to improve the accuracy of weak classifiers. 70% percent accuracy of weak classifiers has been increased using this method.

Golande et al. [7] deal with a system that handles the dataset for both male and female under the age group 25–69 to find the risk of heart diseases. The author elaborates different classification techniques and simply evaluates various accuracy values of different methods.

Gawali et al. [8] uses data extraction techniques to propose a decision support system. The author uses features like blood pressure, gender, cholesterol etc. to forecast the possibility of cardiac infarction. To predict the risk, factor the author performed two techniques like Naïve Bayes and improved K-means clustering and obtained accuracy rate as 84.43%.

UlHaq et al. [9] applied all the classification techniques to design framework that utilizes the dataset Cleveland. The author proposed all the machine learning techniques to find the best approach among them. Among all the methods the support vector machine comes out with highest accuracy as 86%. The least accuracy was given by decision tree and artificial neural network as 74%.

Buchan et al. [10] processes files with Apache technique called Apache cTAKES, which is an NLP based system used for removing details from medical free text. The author has compared a functional removal riddle with PCA and mutual info using three distinct classifiers: (1) Naïve Bayes, (2) MaxEnt and (3) SVM, for comparison purpose randomization testing, is done. Out of all the classifiers, the author was able to get the highest F1 score for the SVM method.

Sowmiya et al. [11] compared classification techniques to examine the prediction of heart problems with the previous framework. Techniques like decision tree, Naïve

Table 1 Represent the literature survey of various researchers

Author	Algorithms/techniques used	Outcome description
Shashikant et al. [3]	Logistic regression Decision tree Random forest	The system design predicts the cardiac failures in smokers. The author applied three algorithms, out of the three algorithms; random forest gives the highest accuracy as 93.61%
Mezzatesta et al. [4]	SVM (Kernel = RBF & Non-linear)	With the use of Support vector machine, the author obtained 95.25% accuracy for Italian datasets and 92.15% for American datasets
Beunza et al. [5]	Decision tree, random forest, support vector machine, neural network and logistic regression	Out of all the supervised machine learning methods, the author gets the highest AUC value 0.75 by support vector machine
Latha et al. [6]	Ensemble techniques like bagging and boosting with other machine learning algorithms	The author achieves 0.7 improved accuracy for weak classifiers
Golande et al. [7]	Decision Tree, KNN, K-means clustering, adaboost	The author able to study about various classification with respective accuracy
Gawali et al. [8]	Naïve Bayes, improved K-means	Two data mining techniques were applied and its accuracy is obtained as 84.43
Ul-Haq et al. [9]	Logistic regression, support vector machine, Naïve Bayes, artificial neural network, decision tree, KNN	By using various algorithms, the author comes to conclusion that support vector machine gives higher accuracy than others and decision tree and ANN gives the least accuracy
Buchan et al. [10]	Naïve Bayes, MaxEnt and SVM	Out of all the classifiers, SVM gives the highest F1 score as 77.4%. Hence, SVM gives the best results
Sowmiya et al. [11]	Decision tree, Naïve Bayesian neural network, ANN, KNN, SVM, apriori algorithm	The author compared all these classification techniques with the existing framework, and found classification technique gives the highest accuracy among all
Karayilan et al. [12]	ANN, back-propagation algorithm	The prediction system able to produce accuracy of 95% using these algorithms

(continued)

Table 1 (continued)

Author	Algorithms/techniques used	Outcome description
Kumar et al. [2]	Multiple logistic regression	The author tried to find which are the most significant variables that indicates the heart problems, and using ROC test, blood sugar (BS) fasting as well as blood sugar levels (BS) post meal enjoy a confident correlation with heart problems
Purushottam et al. [13]	Decision tree	The author converted the database into rules using decision tree, to find the accuracy at different partitions. 86.75% success rate is observed using decision tree
El-Bialy et al. [14]	C4.5, fast decision tree	78.06% accuracy was obtained for collected datasets whereas 75.48% was obtained for separated datasets to predict the risk of CAD problems
Gandhi et al. [15]	Decision tree, neural network, Naïve Bayes	The author analyzed different algorithms to foresee the cardiac infarction at the very first stage
Sreejith et al. [16]	KNN, C4.5, Naïve Bayes, random forest	Reading of Heart rate received after 30-min interval for different person on various parameters

Bayesian neural network, ANN, SVM, KNN etc., among the applied techniques Naïve Bayes gives the better result.

Karayilan et al. [12] developed a healthcare analysis structure influenced by machine learning for prediction of heart issues that offers a lot more proper examination as opposed to the normal manner. In this paper, the writer used the aerobic illness prediction method which uses a man-made neural community back-propagation algorithm. 13 clinical abilities are utilized as responses for your neural telephone system as well as the neural community was instructed together with the back-propagation algorithm to assume the existence or maybe lack of cardiovascular illnesses together with the precision of 95%.

Kumar et al. [2] they model a scalable three-tier structure to the retailer as well as a procedure like a tremendous level of wearable sensor info. Type1 is actually centered on a variety of info offered by IoT wearable sensor gear. Type2 makes use of Apache HBase in keeping the massive level of wearable IoT sensor info found in cloud computing. Furthermore, Type3 utilizes Apache Mahout for enhancing the logistic regression-based prediction type for aerobic illness. Last but not least, ROC is actually carried out to compute one of the most considerable health-related specifics developing amazing issues.

Purushottam et al. [13] implemented a program which can effectively learn the guidelines to predict the amount of risk for heart disease in individual depending on the specified parameters. Certain rules are made which can be prioritized depending on the user's requirement. The functionality of the device examined in terms of the results and classification accuracy that demonstrates the system's potential in forecasting the heart disease risk level much more accurately.

El-Bialy et al. [14] they used the integration on the machine learning method outcomes on distinct datasets concentrating on the CAD illness. This methodology helps to stay away from missing, incorrect as well as inconsistent data issues which could show up in the collected information. Decision Tree and Pruned C4.5 algorithm used to draw a final tree from various datasets and compared. Common features are extracted and analyzed that the accuracy of collected datasets is 78.06% greater as compared to the average of separated datasets which is 75.48%.

Gandhi et al. [15] utilized information mining strategies to foresee heart problems utilizing different algorithms like Decision Tree, Neural Networks as well as Naïve Bayes. The strategies put on to find out concealed patterns to make solutions found health care groups.

Sreejith et al. [16] they proposed a framework for computing the pulse rate, high heat as well as blood pressure levels on the impacted man or women by way of a wearable gadget as well as the estimated parameter is actually transmitted to an android Smart phone empowered with the Bluetooth. The variables are actually examined through the android program on the consumer aspect. The result is actually transferred to the server advantage is actually a regular interval. Each time an immediate scenario comes up aware email is actually forwarded to the various therapy suppliers near the possibility aspect program. The suggested job compares several algorithms as well as proposed the usage of the Random forest algorithm for center trouble prediction.

3 Proposed Methodology

To overcome the data generation using only wearable sensor produces data in large volume which becomes difficult to handle. Providing solution to this IoT has been combined with machine learning. The methodology of machine learning, the dataset which has been used and which classifier gives best accuracy has been explained below:

3.1 Dataset Description

For this proposed system, we have used Cleveland cardiac dataset from UCI research repository. There are 14 properties and 303 occurrences in the dataset. 8 categorical and 6 numerical characteristics are available. Table 2 includes an overview of the

Table 2 Dataset description

S. No.	Attribute name	Description	Range of values
1	Age	Age of person in years	29–79
2	Sex	Gender of the person [1-male, O-female]	0.1
3	Trestbps	Resting blood pressure in mm Hg	94–200
4	Cp	Chest pain type [1-typical type 1 angina, 2-atypical type angina, 3-non-angina pain, 4-asymptomatic]	1, 2, 3, 4
5	Fbs	Fasting blood sugar in mg/dl	0.1
6	Choi	Serum cholesterol in mg/dl	126–564
7	Thalach	Maximum heart rate achieved	71–202
S	Restecg	Resting electrocardiographic results	0, 1, 2
9	Old peak	ST depression induced by exercise relative to rest	1–3
10	Exang	Exercise induced angina	0.1
11	Ca	Number of major vessels coloured by fluoroscopy	0–3
12	Slope	Slope of the peak exercise ST segment	1, 2, 3
13	Thai	3-normal, 6-fixed defect, 7-reversible defect	3, 6, 7
14	Num	Class attribute	0 or 1

dataset. Each study has chosen patients between the ages of 29 and 79. The gender value of the males is 1 and the value for female gender is 0. Symbolic term is given to the four forms of discomfort heart disease. Due to lower blood supply to the shortened coronary arteries, Angina type 1 is triggered. Form 1 Angina is called as an anxiety or mental pressure chest pain. Non-angina chest area soreness could be brought on by a number of indications and sometimes not due to a genuine center encounter. Asymptomatic, your fourth category isn't a cardiac warning sign. Another trestbps attribute will be the measurement of blood pressure levels. Chol would be the quantity of cholesterol. FBS is actually worth for your fast blood sugar levels amount this means whether the value of it is actually one consequently fasting blood sugar levels is actually under 120 mg/dl, as well as zero is actually above. Restecg may be the residual electrocardiographic score; thalach is actually the optimum heart rate. The Angina obtains the value as 1 when pain is present and 0 when no pain occurs. The class rating is 0 for regular and 1 for heart disease patients [6]. The information of dataset is given in Table 2.

3.2 Proposed System

The proposed consist of some procedure which can better explain through dataflow diagram. Figure 4 shows Dataflow diagram of applied methodology. Five methods of

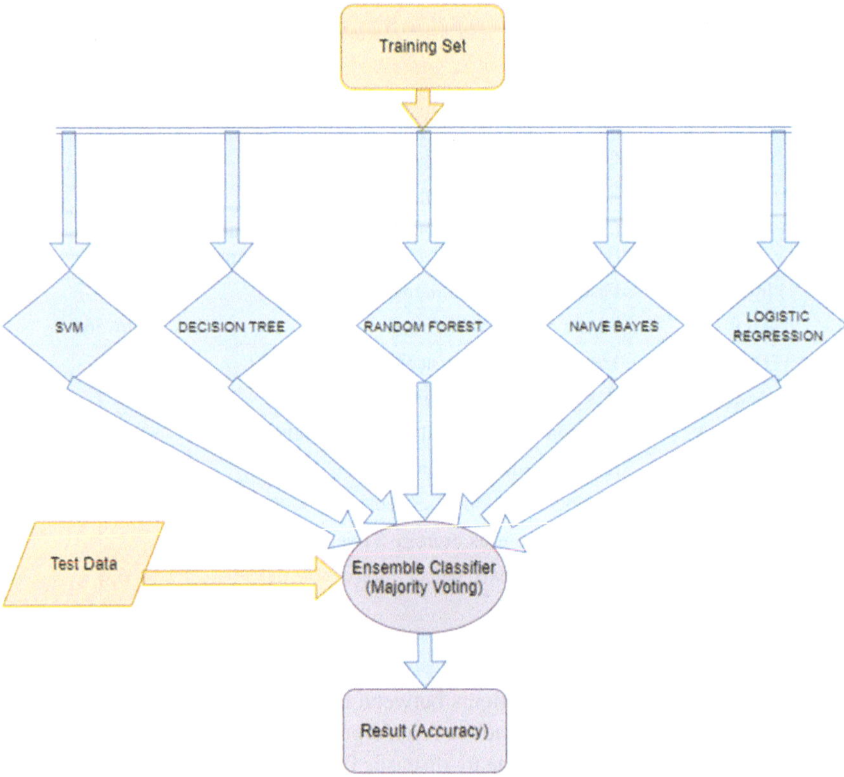

Fig. 4 Dataflow diagram

machine learning have been used to the training set, using majority voting prediction takes place.

4 Classification Techniques

Classification method falls under the supervised learning process which is used to forecast the results based on previous knowledge or data. This paper comes up with a method for classification algorithms for the detection of the heart condition and predicts which algorithms give good accuracy. The data collection is broken into a training sample and a test set. The testing subset is used to train individual classifiers. All the individual classifiers are being elaborated in the next section.

4.1 Naïve Bayes

The description of Naïve Bayes is based on the rule, or simply on that of Bayes Theorem. The Bayes network is a special case, and it is a classifier based on probability. Both roles are conditionally autonomous in the Naïve Bayes network. Consequently, the improvements of one role do not impact another. The software Naïve Bayes can be used to categorize large datasets. The classification algorithm uses the equality of the condition. All the attributes are independent to each other in a set [6].

Let S be a list of learning details and target names. Each row in a dataset has n features defined by $Q = \{B_1, B_2, \ldots B_n\}$. Let target names define by T classes be shown as $(t_1, t_2 \ldots t_n)$. The classifiers determine that X corresponds to a certain tuple Q, which is dependent on Q, with the maximum back likelihood. The method predicts the tuple Q falls to class T_i if and only if

$$P(Ti|Q) > P(Tj|Q) \quad for 1 \leq j \leq T, \quad j \neq i. \tag{1}$$

Hence, out of these two $P(Ti|Q)$ gets the maximum value. The maximum posterior hypothesis is known for the class c_i which belongs to $P(Ti|Q)$. The Bayes theorem states:

$$P(Ti|Q) = \left\{ \frac{P(Q|Ti).P(Ti)}{P(Q)} \right\} \tag{2}$$

The attributes are completely independent to each other [6].

$$P(Q|Ti) = \pi nk = 1P(Qk|Ti) \tag{3}$$

where $Qk = Bk$ for column x.

4.2 Majority Voting

This classifier is actually recognized as a Meta classifier that is used incorporating several classifiers via Majority voting. The last grouping label is going to become the course label which was expected by way of a great bulk of all of the classifiers. Last category label d_j is identified as

$$dj = mode\{C_1, C_2 \ldots, C_3\} \tag{4}$$

where $C_1, C_2 \ldots, C_3$ represents different classifiers.

Majority Voting Algorithm: Let $D_{i,j}$ be the prediction of the ith classifier on a class with j labels.

$$\sum in = 1Di, j = \max j - 1, \ldots m \sum nt = 1Di, j \quad (5)$$

The ensemble classifier's probability for the decision to be better is:

$$Pens = \sum nk = \left(\frac{n}{2}\right) + 1(Cnk)Pk(1 - P)^n - k \quad (6)$$

4.3 Logistic Regression

Logistic regression is actually among the distinctive methods. With this technique, there's a predictive adjustable y in which y € [zero, one]. Right here zero denotes damaging category as well as one for a good category. A multi distinction also is carried out to discover the importance of r when r € [zero, one, two, three]. 2 martial arts classes zero as well as one buy categorized utilizing theory $T(\theta) = \theta T Q$, as well as Tθ(Q) during 0.5, will be the threshold classifier result. When the method predicts r = 1 which suggests the individual is suffering from center disorders, subsequently the theory worth ranges as $T\theta(Q) \geq 0.5$. In the event the theory great is actually under 0.5, then r = 0 that highlights the individual is actually healthful. At last, logistic regression lies between $0 \leq T(\theta) \leq 1$. Sigmod function for logistic regression is followed as [9]:

$$T\theta(q) = G(\theta T Q) \quad (7)$$

where $G(Z)$ *and* $T\theta(q)$ further equals to:

$$G(Z) = \frac{1}{1 + q - Z} \quad and \quad T\theta(Q) = \frac{1}{1 + Q - Z} \quad (8)$$

The cost function for logistic regression can be found as:

$$J(\theta) = 1/m\left(\sum mi = 1cost(T\theta(q(i)), r(i))\right) \quad (9)$$

4.4 SVM (Support Vector Machine)

The SVM is actually a distinction method worn largely for distinction issues. To be able to fix a powerful quadratic programming issue, SVM has spent an optimum margin method. Because of SVM's effective distinction effectiveness, different uses utilized it thoroughly. Within a concern of binary distinction, the cases are actually segregated by way of a $wTx+b = 0$ hyperplane in which w, as well as d, are actually dimensional vectors which are typical to the surface area hyperplane, b stands out as the compensating supply worth, as well as x is actually information established worth. The SVM acquires w as well as blast paper printer. The worth for w may be estimated using Lagrangian multipliers. The boundary information factors are actually viewed as vectors of help (support vectors).The value of w is w = $\sum_i{}^n{}_{=1} \alpha_i y_i x_i$, where n is the count of helper vectors and y_i are the class label to x. The linear discrimination method is as follows [9]:

$$g(x) = sgn\left(\sum in = 1\alpha iyixiTx + b\right) \tag{10}$$

The scenes for nonlinear [9], for decision function and kernel trick obtained by:

$$g(x) = sgn\left(\sum ni = 1\alpha iyiK(xi, x) + b\right) \tag{11}$$

4.5 *Random Forest*

Brieman created the Random forest technique. The strategy of Random forest is actually near to bagging or maybe regression trees with a substantial alter. Rather than snapping just about all possible binary splits on every applicant variable, a person requires binary division in just an arbitrary sample of this applicant predictor variable into account when 1 within one bootstrap analysis creates a distinction or maybe regression tree within a certain node. Just before the technique is specified the dimensions of this group of arbitrarily selected forecast variables. As soon as arbitrary regression tree values become equipped, we put the dimensions of arbitrarily opted predictor variables [p/3], in which p belongs to the entire amount of predictor variables. When distinction tree great is actually equipped, we demand sqrt(p) to always be the dimensions of this pair of arbitrarily selected predictor variables. 500 distinctions, as well as regression trees, has been created. Prediction or perhaps classifications are actually produced by averaging all of the predictions or perhaps by vast majority voting within the distinction trees. All of the characteristics are established to the default of theirs appreciates [17].

4.6 *Cart*

CART [18] stands for Classification and Regression Trees. This process is actually coined by Leo Breiman to affect algorithms of a Decision tree which may be utilized for predictive modeling problems of classification or regression. This method is typically called "decision trees", but it is called a more common word CART on some websites, such as R. CART gives a base for big algorithms such as bagged decision-making bodies, random forests and improved decision trees.

Classification [19] tasks in CART are carried out using the metric Gini index. This metric is addition of squared probabilities of each class [20]. Gini index is calculated as:

$$Gini = 1 - \sum (Pi)2\ for\,i = 1\,to\,n\,classes \tag{12}$$

5 Result Analysis

In order to forecast the risk of heart diseases, five distinct method of machine learning i.e., SVM, Decision Tree (CART), Random Forest, Naïve Bayes, Logistic Regression have been chosen. All the five algorithms have been applied separately and accuracy rate of individual methods are measured as shown in Table 3.

Figure 5 represents the precision rate of existing system. The implemented paper has been compared with the existing method, in exisiting method ensemble techniques like bagging and boosting is applied to machine learning techniques, and through majority voting result was being analyzed as 85.48% accuracy rate as shown in Fig. 5. But the proposed work helps to choose best among the algorithms by the concept of majority voting which is part of ensemble techniques. When multiple output is present, then best one is chosen using majority voting algorithm. On applying majority voting algorithm to these methods, the accuracy rate as 88.59% is obtained. The comparison result is shown in Fig. 6.

In Fig. 6, MV1 and MV2 are the algorithms used in exiting system and MV3 is the algorithm used in proposed system. Therefore, the proposed system gives the best result than previous one.

Table 3 Accuracy of different applied algorithms

Algorithms	Accuracy (in percentage)
Support vector machine	87.995
Decision tree (CART)	88.004
Random forest	89.965
Naïve Bayes	89.057
Logistic regression	92.009

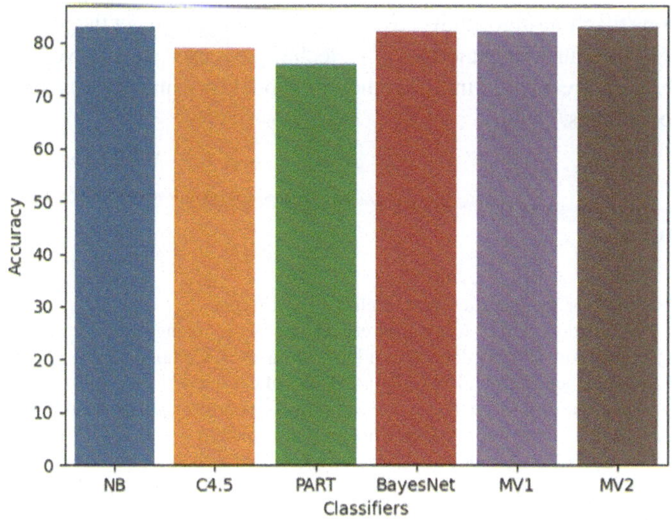

Fig. 5 Result of existing system

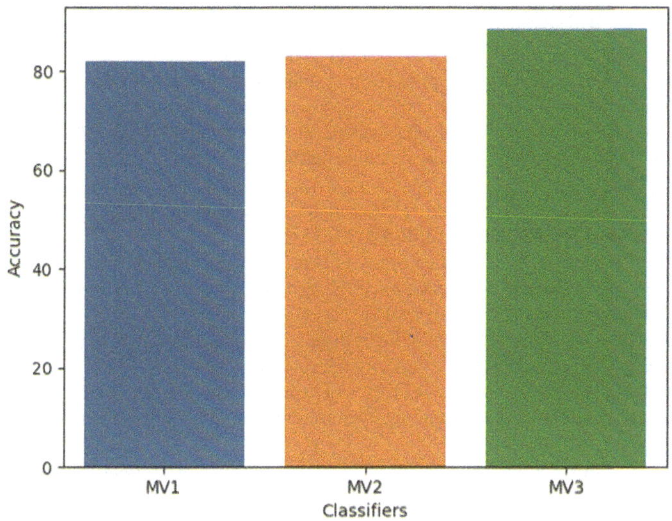

Fig. 6 Comparison of existing and proposed system

6 Conclusion

In this paper, the machine learning is used to handle the volume of data produced using IoMT. The precision of predicting the heart problems is implemented using machine learning algorithms. Logistic regression gives the highest accuracy which is

92%. Various machine algorithms have been implemented to get the most promising accuracy. For this purpose, majority voting technique is used which gives the accuracy of 88.59%. In future we will try to predict the model by using various different deep learning techniques.

References

1. https://www.medicalnewstoday.com/articles/237191#types
2. Kumar, P.M., Gandhi, U.D.: A novel three-tier Internet of Things architecture with machine learning algorithms for early detection of heart diseases. J. Comput. Electr. Eng. **65**, 222–235
3. Shashikant, R., Chetankumar, P.: Predictive model of cardiac arrest in smokers using machine learning technique based on heart rate variability parameter. Appl. Comput. Inf. (2019). ISSN:2210-8327. https://doi.org/10.1016/j.aci.2019.06.002n
4. Mezzatesta, S., et al.: A machine learning-based approach for predicting the outbreak of cardiovascular diseases in patient on dialysis. Comput. Methods Prog. Biomed. **177**, 9–15 (2019)
5. Beunza, J.-J., et al.: Comparison of machine learning algorithms for clinical event prediction (risk of coronary heart disease). J. Biomed. Inf. **97**, 103257 (2019)
6. Beulah Christalin Latha, C., Carolin Jeeva, S.: Improving the accuracy of prediction of heart disease risk based on ensemble classification techniques. Inf. Med. Unlocked **16**, 100203 (2019)
7. Golande, A., Pavankumar, T.: Heart disease prediction using effective machine learning techniques. Int. J. Recent Technol. Eng. (IJRTE) **08** (2019)
8. Gawali, M., et al.: Heart disease prediction system using data mining techniques. Int. J. Pure Appl. Math. **120**, 499–506 (2018)
9. UlHaq, A., Ping Li, J., Hammad Memon, M., et al.: A hybrid Intelligent system framework for prediction of heart disease using machine learning algorithms. Mobile Inf. Syst. Article ID 3860146 **2018** (2018)
10. Buchan, K., Filannio, M., Uzuner, O.: Automatic prediction of coronary artery disease from clinical narratives. J. Biomed. Inform. **72**, 23–32 (2017)
11. Sowmiya, C., Sumitra, P.: Analytical study of heart disease diagnosis using classification techniques. In: International Conference on Intelligent Techniques in Control, Optimization and Signal Processing (pp. 1–5). IEEE (2017)
12. Karayilan, T., Kilic, O.: Prediction of heart disease using neural network. In: 2nd International Conference on Computer Science and Engineering, pp. 719–723 (2017)
13. Purushottam, Saxena, K., Sharma, R.: Efficient heart diseases prediction using decision tree. In: International Conference on Computing, Communication and Automation (ICCCA), pp. 72–77 (2015)
14. El-Bialy, R., Mostafa Salamay, A., et al.: Feature analysis of coronary artery heart disease datasets. In: International Conference on Communication, Management and Information Technology (ICCMIT), vol 65, pp. 459–468 (2015)
15. Gandhi, M., Narayan Singh, S.: Prediction in heart disease using techniques of data mining. In: 1st International Conference on Futuristic trend in Computational Analysis and Knowledge Management, pp. 520–525 (2015)
16. Sreejith, S., Rahul, S., Jisha, R.C: A real time patient monitoring system for heart disease prediction using random forest algorithm. J. Adv. Sig. Process. Intell. Recogn. Sys. **425**, pp. 485–500 (2015)
17. Peter Austin, C., Jack Tu, V., Jennifer Ho, E., et al.: Using methods from the data-mining and machine learning literature for disease classification and prediction: a case study examining classification of heart failure subtypes. J. Clin. Epidemiol. **66**, 398–407 (2013)
18. https://machinelearningmastery.com/classification-and-regression-trees-for-machine-learning/

19. https://sefiks.com/2018/08/27/a-step-by-step-cart-decision-tree-example/
20. Khamparia, A., Pande, S., Gupta, D., Khanna, A., Sangaiah, A.K.: Multi level framework for anamoly detection in social networking. Library Hi-Tech 10.1108/LHT-01-2019-0023

QoS Optimization in Internet of Medical Things for Sustainable Management

Ashu Gautam, Rashima Mahajan, and Sherin Zafar

Abstract The Internet of things (IoT) technology brings a significant revolution traditional healthcare sector. However, the amalgamation of IoT with medical equipment, also widely known as internet of medical things (IoMT), gave numerous susceptibilities at different layers of modern healthcare sector. The field of IoMT is increasing because of the various advantages associated mainly for detection, prevention and diagnosis of numerous diseases in patients. The risks associated with the security of IoMT, still needs to be addressed as in the core of this new technology, Wireless Sensor Networks (WSN) along with Mobile Ad-hoc Network (MANET) are most widely employed for picking information physically from environment by the IoT for better user friendly experience. This interaction used in various systems is famously called as MANET-IOT systems and IT based networks. The comparison has been made between reactive, secured and hybrid protocols for the various quality of service (QoS) parameters under the scenario of changing number of connections. The result emphasized that Hybrid Wireless Mesh Networks (HWMP) performed more efficiently when compared with Ad-hoc on Demand Distance Vector (AODV) and Secured AODV (SAODV) routing protocol.

Keywords IoT-MANET · AODV · SAODV · HWMP · Adhoc networks for E-healthcare

A. Gautam (✉) · R. Mahajan
MRIIRS, Faridabad 121004, India
e-mail: ashuone@gmail.com

R. Mahajan
e-mail: rashima.fet@mriu.edu.in

S. Zafar
Department of CSE, SEST, Jamia Hamdard, New Delhi 110062, India
e-mail: zafarsherin@gmail.com

A. E. Hassanien et al. (eds.), *Cognitive Internet of Medical Things for Smart Healthcare*, Studies in Systems, Decision and Control 311,
https://doi.org/10.1007/978-3-030-55833-8_10

1 Introduction

Over the last decade, medical equipment and their applications connected through WMN and mobile computing fields are undergoing dynamic developments. The IoT and its amalgamation with medical field related devices play the important role in the Internet of Medical Things (IoMT). For home assisted health monitoring of old age persons, patients who are unable to bear the expenses of hospitals, various researchers, practitioners and in the conditions like that of COVID-19 where there is a restriction in movement of patients the IoMT, seems to be an answer of every concern. Thus it is very much expected to have an tremendous growth in the demand of IoT in medical field by 2025. One of the research emphasized that IoT based healthcare market is dignified to reach $117 by end of 2020 [1, 2]. This is credited to the associated advantages of IoMT and wireless mesh networks (WMN), majorly because of easy scaling; rapid deploy ability and the most significant cost-effective intelligent devices for a better smart world. Figure 1 shows the medical devices and the important role they play when connected to each other in a network, forming mesh network. The major functions highlighted are related to diagnosis of disease, also their involvement in prevention of disease by providing some alert to the user and finally IoMT devices could facilitate in providing the treatment to the patients based on the utilization of technologies like artificial intelligence (AI) and machine learning (ML) [3, 4].

The services offered by IoT in healthcare sector based on WMN vary from a simple task of maintenance of physical and electronic health record known as PHR,EHR respectively of a particular patient comprising its basic information (name, age, gender, address etc.) to a critical task of reducing or even completely eliminating the symptoms of several severe critical condition or stage of asthma, cancer, diabetes and also alarming heart conditions, which can lead to dangerous situation in patients life

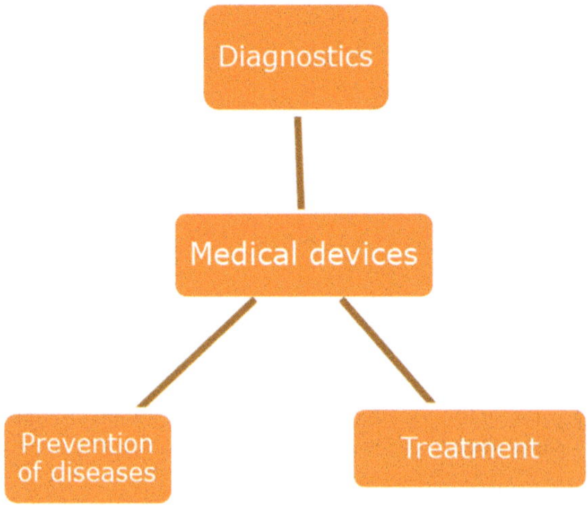

Fig. 1 Role of medical devices in IoMT

by sharing the vital characteristics of patients from on device or application to other data centers situated all over the world. Irrespective of range, all types of networks allow computers along with entities to share data and assets among each other [5, 6].

Since the applications based on IoT in healthcare sector are mainly related to handle patients data in form of sending information from one medical devices (invasive, non-invasive or active) to other sensing device which may be far off their radio range. Thus, routing is a very crucial facility for the success of end-to-end announcement. Therefore, this amalgamation of IoT in sector of healthcare is also responsible for various new attacks and threats in the devices responsible for sending and receiving vital information in healthcare area. One of the main reason for presence of different types of attacks is the heterogeneous medical instruments involved in working of healthcare sectors ultimately giving complexity and incompatibility in working operation. Secondly different manufacturer, involved in make of medical devices are not concerned about the security and privacy part of devices involved in IoMT, thus issues like integrity along with confidentiality are prevailing. All the violations in security related to WSN are automatically forming the part of IoMT based healthcare applications, since all the devices engaged are sending data wirelessly. So far, many reactive and hybrid routing protocols have been suggested for networks. An attacker can insert malevolent routing data into network, causing in contradictions in the routing and thus can be danger to patient's life. Mobile Ad hoc Network (MANET) has ability of creating network without some set-up for an emergency or for a temporary period [5, 6]. Therefore because of this convergence of MANET and WSN, MANET is appropriate to be integrated with internet of things (IoT) for healthcare environment.

1.1 MANET-IoT Based Smart Healthcare Networks

This section elaborates the IoMT based health care system architecture, broadly consisting of layers as shown in Fig. 2 for remote monitoring of health. The first layer is known as End point layer which in turn consists of different types of medical devices used in transmitting the information from one point to other, one type includes wearables medical devices for example the 12 lead configuration for electrocardiogram (ECG), electromyogram (EMG), electroencephalogram (EEG)etc. Another type are stationary medical devices for example the CT scan machines, third category is ambient devices which picks the information from the surrounding of the patients environments for example the cameras involved in image processing and the last set are implantable devices for example the programmable pacemakers which are internally implanted in the patient's body [7]. Second layer is called as gateway through which devices connect to each other; comprised of routers, switches and firewall. The third component of the IoMT landscape is the technology used for handling the patients data; the most popular platforms could utilize data analytics science, web interface or may have the backup storage. Typically this layer collects the data and also integrates the same with external data and incorporates both local processing

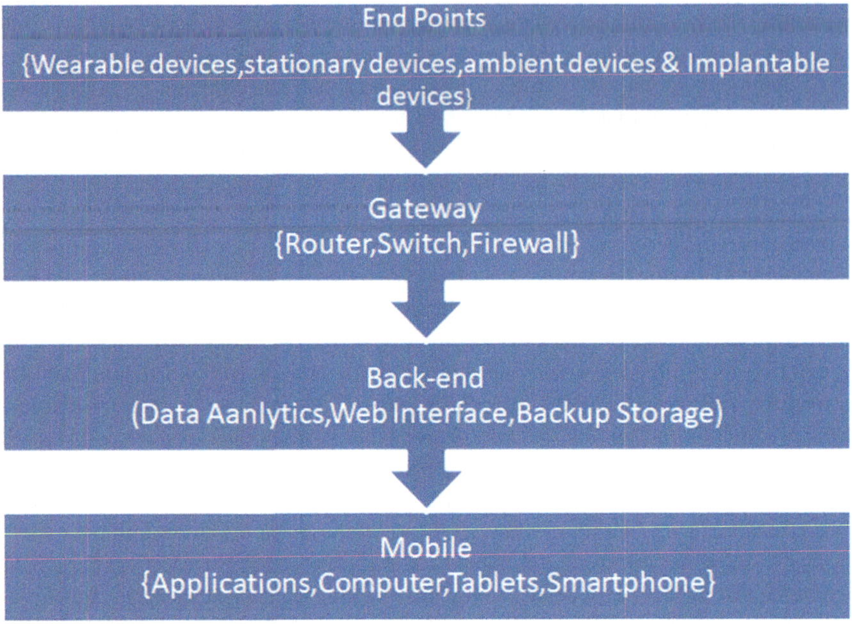

Fig. 2 IoMT components landscape in healthcare

analysis and storage of data as well as complex processing. The fourth and the last layer is quiet mobile in nature comprising the smartphones, tablets, laptops etc. where applications are installed [8, 9].

This cloud layer is very much susceptible to many kinds of attacks; ultimately resulting in denial-of-service (DDOS) at each layer [10]. The routing decision layer comprise of IoT devices and sensor controller unit. The IoT devices have a direct link with the local processing, storage and analysis layer via Wi-Fi, GSM, Ethernet 2G/3G etc. The smart phone could be an IoT device, which everyone is using for monitoring various health parameters via different application provided by healthcare providers. Thus this IoT—mobile adhoc network (MANET) amalgamation is widely used for monitoring, detection and prevention of disease in the healthcare sector. These sub-layers are capable, for processing sensor output and retrieving context, integrating with cloud and providing feedback to and from the patient [11, 12].

Patients health related data is continuously engendered from various apps which we have on various devices like smartphones, desktop, laptops, and wearables gadgets, connected by various communication technologies like Wi-Fi, Internet, and Bluetooth etc., engaged not only in monitoring or tracking our habits but also giving us feedback for better health related issues. This massive amount of huge data which is ever increasing on daily basis is termed as Big Data [13]. The smart world concept is feasible only by implementing IoT in various domains like healthcare sector along with the AI and ML techniques. IoMT based technology associates different intelligent devices, for exchanging patients health related data [14].

The various advanced techniques used in manufacturing of medical devices made them compact, cheaper, light in weight. Also, the resource allocation can be utilized in better way, in patients care. The wide range of IoMT devices comprise of cconsumer health-monitoring technology (in form of watches, health bands, smart watches etc.), internally implanted-embedded medical devices (such as programmable pace-makers for functioning of heart), externally placed wearables devices (fibrillators, gas analyser etc.), stationary medical device [15].

However, all these devices used in healthcare sector connected to each other via unsecured network environment is ultimately bringing a security risk in basic operation and thus bringing harm to patient's health and associated security. Further these IoMT devices are manufactured without consideration of security aspects which makes them vulnerable to various kind of attacks. One of the report on healthcare highlights that the propagation of IoMT based on IoT devices for healthcare, alongside with absence of network segmentation, inadequate admittance controls and dependence on legacy systems, has led to an increased surface for attacks which can be misused by malicious nodes or attackers dogged to snip personal identification information (PII) and protected health records (PHR), in addition to distracting delivery operations of healthcare [16].

1.2 Assaults in WSN and IoT Based Networks

The relevant analysis for the different types of attacks faced by wireless sensor networks reflects that the attacks are broadly classified as active attacks and passive attacks category respectively. Another subcategory much talked about in literature is based on the description of the layered stacked protocol of open systems interconnection (OSI) model. Thus active attacks could further be categorized on the layers of OSI, attacks may vary at different layers physical layer, medium access control (MAC), network layer, transport layer and application layer [17]. The exposure to the internet is making world small undoubtedly. However, the exposure to the internet also invites attackers to invade into the system thus confidentiality, privacy of data are at stake. The remote monitoring of patients health especially in the crucial phase of coronavirus is successful and possible only in the light of secure transmission of data from remote locations. In this research study we are considering the most serious assault which is DDoS attack which can hamper the usage of the resources for indefinite time to the users. In this research work syn flood attacks are implemented in NS2 for evaluating the performances of most widely used routing protocols in MANET based environment. The topmost concern as regards to syn flood attack is flooding of complete network by the malicious device also called as node. Thus the devices such as smartphones, laptops, smart wearables devices like watches, desktops which are many time depended on battery usage will get impacted by such attacks as syn flood attacks will consume their bandwidth and energy ultimately resulting into performance degradation of the networks [18–20].

Introduction section of the research paper has focussed on concepts of merger IoT with medical devices in healthcare sector and the security challenges faced by IoMT, further in Sect. 2 relevant quality of service (QoS) of routing protocols significant for adoption in WMN based networks has been discussed for tackling security in healthcare sector based on WMN and MANET. Section 3 depicts broad area that focuses on various protocols of used in transmission layer. Section 4 includes the scenario of simulation model used in this research work followed by results. The performance of three routing protocols AODV, SAODV and HWMP in Adhoc MANET scenario of changing no. of connections considering throughput, PDR (packet delivery ratio) NRL (Normalized Load Ratio), throughput, delay and total energy consumption as evaluating criteria's for Quality of Service (QoS). The discussions and conclusion on the research work conducted are listed in Sect. 5, followed by future and current work of research and references are listed in the end.

2 QoS for WMN Based Smart Infrastructure

Clinics are important supporters of common asset consumption and natural change. The energy consumption in hospitals buildings and specifically in the field of telemedicine among different layers needs attention from the prospect of developing sustainable health care sector [21]. In the research work to evaluate the QoS of three routing protocols the various metrics of performance considered are delay, packet delivery ratio (PDR), normalized routing load (NRL), throughput and total energy consumption. The particulars of considered metrics are described as:

PDR: The ratio of the number of packets received by the target node to the number of packets actually sent by the source node is called PDR. End-to-End delay: The average time for a packet successfully transmit a message from the source to the destination across the network. Throughput: Throughput refers to the number of bits received by the destination node during simulation time. NRL: The normalized routing load is the ratio of the routing control packet to the number of packets received by the destination node. Routing control packets consume the available bandwidth and battery power of the nodes. Energy consumption is the steady quality; this metric is measured as the ratio of total energy consumed by each node in the network divided by its initial energy. The initial and final energy left in the node are measured at the end of the simulation run. The various sensors which are connected wirelessly, also commonly known as nodes, rely mostly on low powered batteries for their working operation. Thus, suitable design optimization of such systems is mainly the maintenance of this energy so that system incorporating such sensor nodes could work lengthier. A mobile wireless sensor network is additionally susceptible to security risks due to the dynamic in and out movement of modes in various different directions at different pace in comparison to other networks. This also makes them more prone to the various types of attacks most predominantly distributed denial of service attack (DDoS) [22]. The goals of reducing the energy consumed in such nodes can be accomplished in both active and inactive communication held among them. The scenario in which all the existing nodes are participating in exchanging the data is categorized as active communication. Thus to minimize energy in active communication is possible via two different approaches one is transmission power control and second is load distribution. While during inactive communication all the nodes are sitting idle. Therefore in such case, to diminish energy consumption power/sleep down methodology is widely used. Jitter is the variation of data communication packets in the network.

3 WSN—MANET Based Routing Protocols

The wireless sensor networks and the collaboration of mobile adhoc networks with influence of IoT, has huge number of applications in various fields of emergency situations, where remote monitoring is extremely difficult. Like the inability to communicate around disaster location of earthquake or even the remote monitoring of patients health in the serious situation like caused by the COVID-19. WSN have efficient way to send the data from source device to the destination node with the help of routing protocols. Broadly based on the transmission mode these protocols, engaged in finding the best route of path for the data packets, could be classifies and reactive protocols, proactive protocols and the combination of previously mentioned protocols known as hybrid protocols [23].

The researchers have shown interest in the utilization of AODV as reactive protocol also called as on demand driven type of protocol, used in wireless adhoc networks. The key components of the main principle of this reactive routing protocol are categorized as route discovery phase and other as route maintenance phase of AODV as shown in Fig. 3. It utilizes sequence number, route request (RREQ), route reply (RREP). There are few universal performance metrics used along with parameters related to the network environment [24, 25].

AODV requires time to complete the routing table and with the increase in the size of network there is decrease in the generalized metrics of performance. Also due to the breakage of links among nodes large number of packets is sent in AODV thus making congestion in network.

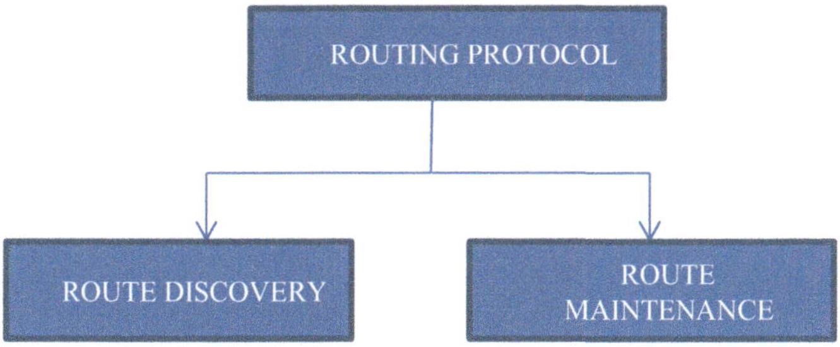

Fig. 3 Routing protocols principle components

The second routing protocol considered in this research work is SAODV. The major dissimilarity between Secured Adhoc On Demand Vector (SAODV) and AODV them is the routing discovery process, SAODV utilize exchanging of some random numbers to verify the target node or device. SAODV enhances an extension to altered AODV packet setups to include digital signature aimed at shielding the non-mutable data. Also hash chains to shield the mutable information (hop count) [26, 27].

Hybrid have both word stands for combination, hybrid protocol thus possess bot the features of proactive and reactive routing protocol. The WMN is using one such hybrid protocol known as hybrid wireless mesh protocol (HWMP) as a by default protocol for mesh networks as per the IEEE standard. HWMP is possesses the following characteristics.

- It is based on AODV and tree based routing technique.
- HWMP uses MAC address instead of IP addresses
- Path selection happens with path election protocol, which could be reactive (on demand) and proactive (tree building mode).On demand scenario: Mesh point (MP) broadcast 'path request'. Destination MP in response to it send unicast path reply.
- It has four elements route announcement (RANN), route request (RREQ), route reply (RREP) and route error (RERR). Among these RANN, RERR are broadcast, while RREP is unicast and RREP is both unicast as well as broadcast [28, 29].

For Proactive Scenario: The MP called as source MP can initiate discovering route process in two ways. Firstly, it announce its presence by periodically sending the root announcement (RANN), that disseminates metric information across the network. When MP receives RANN it creates a path to the root MP and sends PREQ to the root MP, it than unicast PREP to each PREQ. Secondly root MP proactively disseminate PREQ to all MP.MP unicast PREP to root MP, provided PREQ contains a greater sequence number (SN). All the MP maintains unique destination sequence number which makes HWMP loop free like AODV characteristics. Mesh points monitors all the upstream links and can switch back to any links with the help of other element RREP thus avoiding the reconstructing of tree. The RREP helps nodes to choose their own back up routes. The pseudo code is given below for path selection

```
Algorithm: Path selection using HWMP
Input:'n' number of nodes
      'RREQ'  Route Request packet
      'RREP'  Route Reply packet
      'RANN' Route Announcements
Start
If proactive
        Intiate root MP
        Broadcast RANN
       If  RANN received by MP
              Refresh root
              Send PREQ to root MP
              If PREQ received by root MP
              Unicast PREP to each MP
      else
           forward PREQ o next node
       end
       end
else if reactive
  source ——> broadcast PREQ
if PREQ received by intermediate node
forward path
else if PREQ received by Destination node
     accept packet
     generate PREP
     unicast PREP to source
  end
  end
end.
```

The ZRP, Hybrid Clustering Routing (HCR) and Ant based Hybrid routing algorithm for mobile and adhoc networks (ANTHOCNET) [29–31] have been reported in various researches for discovering routes. Additional originality of hybrid routing procedures is that to establish the network in accord with network parameters.

4 Simulation Results

This research study is implemented on mobility model for evaluating AODV, SAODV, HWMP routing protocols. The important characteristic of Adhoc networks is the frequent changes in incoming and outgoing nodes for sending the information from source to destination. Simulation environment setup for mobility based model with varying number of connection for fixed 30 nodes considered on three protocols AODV, SAODV, HWMP MANET is shown in Table 1.

Table 1 Simulation environment with changing number of nodes on routing protocols

Parameters	Value used
No. of nodes	30
Area traffic	1800 × 840 CBR
Simulation time	60 s
No. of connections Traffic rate Speed	5, 10, 15, 20, 25 4 packets/s 20 m/s
Packet size	1024

Table 2 Assessment of PDR with changing number of connections for routing protocols

Routing protocols	PDR for varying no. of connections (5, 10, 15, 20, 25)				
	5	10	15	20	25
AODV without attack	45.0836	47.183	50.921	52.9782	59.8371
AODV with attack	39.8293	41.0381	43.8521	45.1782	51.049
SAODV without attack	59.8372	62.9736	67.948	69.431	74.819
SAODV with attack	54.9274	58.2732	65.0923	67.276	73.082
HWMP without attack	64.379	67.0842	71.7462	72.303	78.9274
HWMP with attack	59.0274	61.0472	67.0384	68.645	74.0947

The simulation results for AODV, SAODV and HWMP are presented by varying no. of connections for QoS parameters namely PDR, NRL, throughput, delay and energy consumption are presented. Comparison of PDR for three routing protocols is shown in Table 2. Figure 4 represents the bar graph.

PDR is high both under attack and in absence of attacks in HWMP in comparison to two other routing protocols.

Comparison of NRL for three routing protocols is shown in Table 3. Figure 5 represents the bar graph for the same. The cost of control overhead for each data packet sent in the network is high for HWMP under both conditions of presence and absence of attacks.

Comparison of end to end delay for three routing protocols is shown in Table 4. Figure 6 represents the bar graph for the same. The delay is transmitting the data is in case of hybrid routing protocol (HWMP)in comparison to other two routing protocols under both conditions of presence and absence of attacks.

Comparison of throughput for three routing protocols is shown in Table 5. Figure 7 represents the bar graph for the same. The throughput is highest in HWMP when compared with other two protocols under both conditions of presence and absence of attacks as we increase the number of connections in scenario considered.

Comparison of energy consumption for three routing protocols is shown in Table 6. Figure 8 represents the bar graph for the same. The energy consumption is less

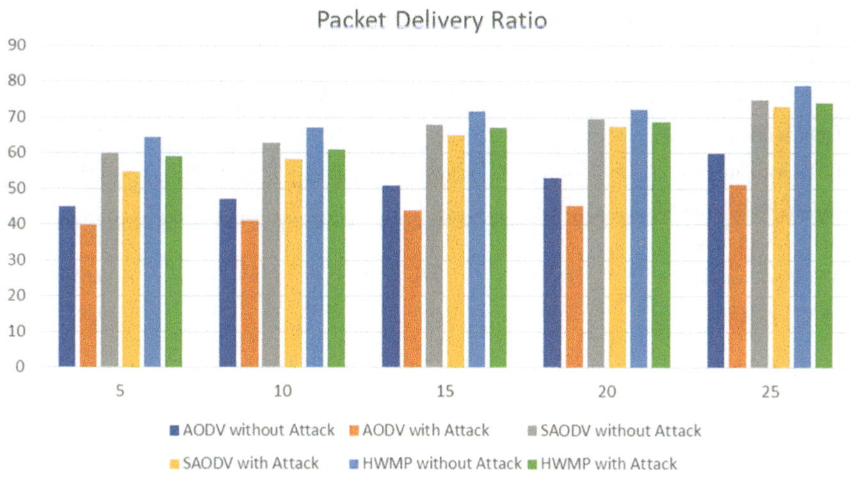

Fig. 4 Comparison of PDR for routing protocols w.r.t number of connections

Table 3 Assessment of NRL with changing number of connections for routing protocols

Routing protocols	NRL for varying no. of connections (5, 10, 15, 20, 25)				
	5	10	15	20	25
AODV without attack	1.65	1.93	2.01	2.17	2.83
AODV with attack	2.974	3.228	3.732	3.906	4.132
SAODV without attack	1.893	2.103	2.561	2.917	3.104
SAODV with attack	3.402	3.827	4.19	4.813	4.918
HWMP without attack	2.318	2.572	2.894	3.016	3.412
HWMP with attack	4.108	4.519	4.903	5.357	5.483

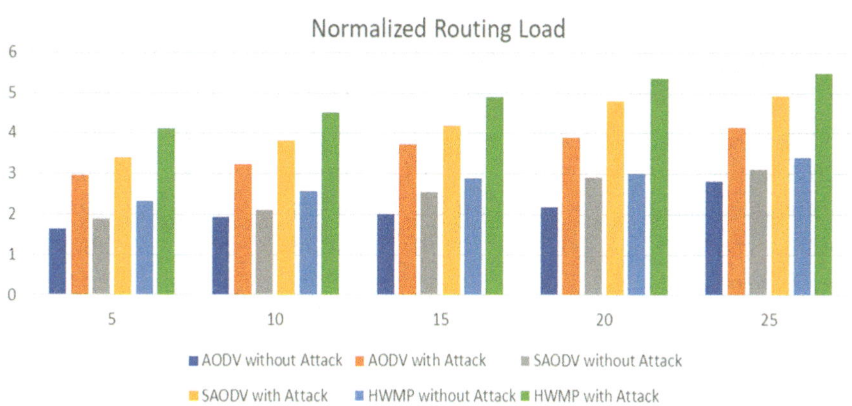

Fig. 5 Comparison of NRL for routing protocols w.r.t number of connections

Table 4 Assessment of delay with changing number of connections for routing protocols

Routing protocols	Delay for varying no. of connections (5, 10, 15, 20, 25)				
	5	10	15	20	25
AODV without attack	0.48	0.53	0.57	0.63	0.71
AODV with attack	0.81	0.87	0.91	0.98	1.09
SAODV without attack	0.44	0.49	0.54	0.61	0.68
SAODV with attack	0.69	0.72	0.78	0.85	0.94
HWMP without attack	0.39	0.43	0.48	0.61	0.65
HWMP with attack	0.61	0.67	0.72	0.83	0.89

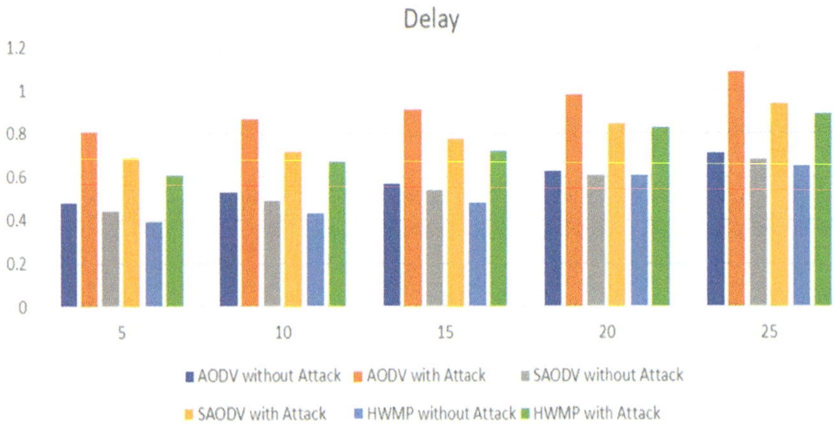

Fig. 6 Comparison of delay for routing protocols w.r.t number of connections

Table 5 Assessment of throughput with changing number of connections for routing protocols

Routing protocols	Throughput for varying no. of connections (5, 10, 15, 20, 25)				
	5	10	15	20	25
AODV without attack	168.28	175.49	182.03	193.92	201.74
AODV with attack	133.74	145.97	159.28	164.83	170.67
SAODV without attack	183.92	194.81	203.75	221.45	239.02
SAODV with attack	151.68	159.04	167.73	179.1	191.93
HWMP without attack	210.82	233.07	259.9	276.36	291.64
HWMP with attack	181.46	188.6	195.27	201.82	229.71

in HWMP when compared with other two protocols under both conditions of presence and absence of attacks as we increase the number of connections in scenario considered.

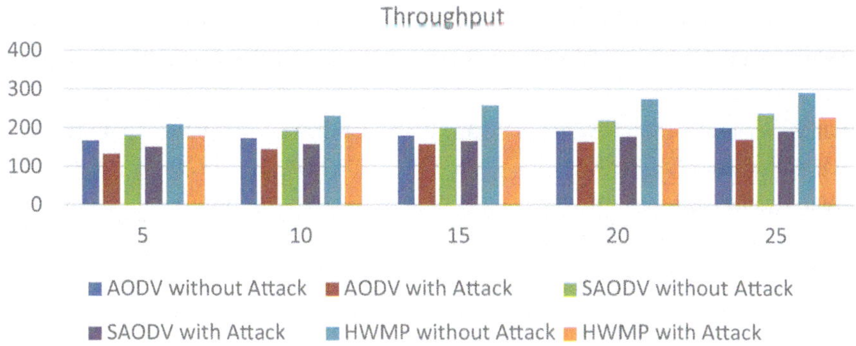

Fig. 7 Comparison of throughput for routing protocols w.r.t number of connections

Table 6 Assessment of energy consumption with changing number of connections for routing protocols

Routing protocols	Throughput for varying no. of connections (5, 10, 15, 20, 25)				
	5	10	15	20	25
AODV without attack	1876.9	1901.43	1989.81	2155.07	2209.63
AODV with attack	1998.73	2118.28	2376.93	2564.53	2883.71
SAODV without attack	1718.9	1794.03	1836.62	2091.54	2183.82
SAODV with attack	1972.86	2009.38	2190.8	2256.9	2472.3
HWMP without attack	999.43	1008.52	1021.7	1099.35	1208.73
HWMP with attack	1872.61	1997.03	2076.54	2109.82	2263.8

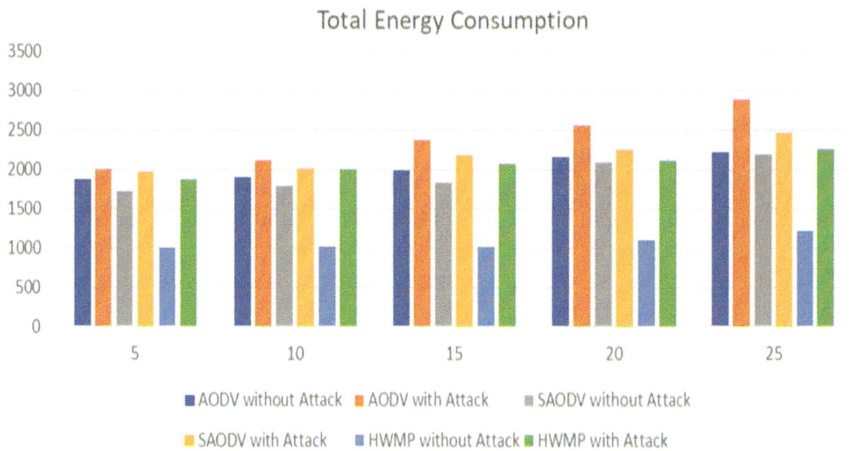

Fig. 8 Comparison of energy consumption for routing protocols w.r.t number of connections

Table 7 Assessment of packet loss ratio with changing number of connections for routing protocols

Routing protocols	Packet loss ratio for varying no. of connections (5, 10, 15, 20, 25)				
	5	10	15	20	25
AODV without attack	54.9164	52.817	49.079	47.0218	40.1629
AODV with attack	60.1707	58.9619	56.1479	54.8218	48.951
SAODV without attack	40.1628	37.0264	32.052	30.569	25.181
SAODV with attack	45.0726	41.7268	34.9077	32.724	26.918
HWMP without attack	35.621	32.9158	28.2538	27.697	21.0726
HWMP with attack	40.9726	38.9528	32.9616	31.355	25.9053

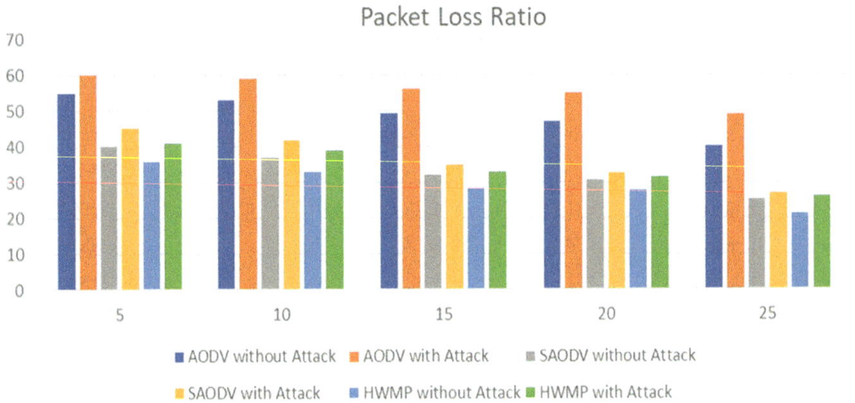

Fig. 9 Comparison of packet loss ratio for routing protocols w.r.t number of connections

Assessment of PLR for three routing protocols is shown in Table 7. Figure 9 represents bar graph shows PLR variation for three protocols in absence and in presence of attacks respectively.

Thus HWMP comparatively provides better results than AODV and SAODV hence, HWMP can be quiet beneficial to be adopted for the healthcare environment.

5 Discussions and Conclusion

For ensuring privacy and security of patients data during transmission in healthcare sector MANET could be used as one of the important technology combined with IoT. The utilization of smart devices for monitoring the health of patients in remote locations for tackling coronavirus is the need of every developing country. The research study carried could be applied in enhancing the most suitable routing protocol for remote monitoring of patients health. For dealing with end to end data transmission of patients sensitive data MANET protocol play a vital role for sending the information securely.This research work analyzed the performance of three routing protocols AODV, SAODV and HWMP in absence of DDoS attacks and under the sway of DDoS attacks. The simulation results of NS2 for the two QoS parameters evaluates that HWMP outperforms from the other two traditional routing protocols for varying number of nodes considered in the simulation. The effect of DDoS attacks have been shown and compared. It is concluded for packet delivery ratio as QoS HWMP is 24.18 and 5.20% better than AODV and SAODV in absence of impact of attacks for maximum connections, under the impact HWMP is 31.10 and 1.3% better than AODV and SAODV respectively. For NRL, HWMP is 17 and 9% better than AODV and SAODV in absence of impact of attacks for maximum connections, whereas under the impact HWMP is 24.63 and 10% better than AODV and SAODV respectively. For delay the HWMP is 9.2 and 4.6% better than AODV and SAODV in absence of impact of attacks for maximum connections, under the impact HWMP is 22 and 5.6% better than AODV and SAODV respectively. HWMP shows that the throughput is also maximum in case of hybrid protocol considered, HWMP is 30.8 and 18.4% better than AODV and SAODV in absence of impact of attacks for maximum connections, under the impact HWMP is 25.70% and 16.49% better than AODV and SAODV respectively The energy consumption at maximum connection HWMP consumes is 82.8 and 80.6% less energy than AODV and SAODV in absence of impact of attacks for maximum connections, under the impact HWMP consumes 27.38% and 9.2% better than AODV and SAODV respectively. The packet loss ratio is also less in HWMP, 90.5% and 19.4% than AODV and SAODV respectively in absence of attacks. Under the influence of attacks for same scenario HWMP is 88.9% and 3.9% better than AODV and SAODV respectively for the loss in packets. The result emphasize that hybrid routing protocol should be preferred in situations where data to be transmitted is very important and any packet loss during transmission is not at all desirable specifically related to healthcare sector.

6 Future Developments

In future, this research study will be considering the measures to works on enhancement of security feature of HWMP routing protocols as security being the primary concern in all the healthcare related information exchange of patients. Thus to enhance security block chain will be utilized to ensure the legitimated smart devices, for monitoring vital characteristics of patients, which is the key issue in E-health care data transfer.

References

1. Agrawal V, (2014) Security and privacy issues in wireless sensor networks for healthcare. In International Internet of Things Summit Springer, Cham pp:223–228 https://link.springer.

com/chapter/10.1007/978-3-319-19656-5_32.
2. Alippi, C., D'Alto, V., Falchetto, M., Pau, D., Roveri, M.: (2017) Detecting changes at the sensor level in cyber-physical systems: Methodology and technological implementation. In 2017 International Joint Conference on Neural Networks (IJCNN) IEEE, pp: 1780–1786. https://ieeexplore.ieee.org/abstract/document/7966066.
3. Alvarez J, Maag S, Zaïdi F(2017) Monitoring dynamic mobile ad-hoc networks: A fully Distributed Hybrid Architecture. In IEEE 31st International Conference on Advanced Information Networking and Applications (AINA) IEEE pp: 407–414. https://ieeexplore.ieee.org/abstract/document/7920938
4. Brill, C., Nash, T.: A comparative analysis of MANET routing protocols through simulation. In: 12th International Conference for Internet Technology and Secured Transactions (ICITST), IEEE pp: 244–247 (2017). https://ieeexplore.ieee.org/abstract/document/8356392
5. Chelli, K.: Security issues in wireless sensor networks: attacks and countermeasures. In: Proceedings of the World Congress on Engineering, 1, 20 (2015). https://www.iaeng.org/publication/WCE2015/WCE2015_pp519-524.pdf
6. Degirmencioglu, A., Erdogan, H.T., Mizani, M.A., Yılmaz, O.: A classification approach for adaptive mitigation of SYN flood attacks: preventing performance loss due to SYN flood attacks. In: NOMS IEEE/IFIP Network Operations and Management Symposium pp. 1109–1112 (2016). https://ieeexplore.ieee.org/abstract/document/7502971
7. Džaferović, E., Sokol, A., Abd Almisreb, A., Norzeli, S.M.: DoS and DDoS vulnerability of IoT: a review. Sustainable engineering and innovation, 1, 1 (2019). https://sei.ardascience.com/index.php/journal/article/view/36
8. Fu, Y., Ding, Z.: Hybrid channel access with CSMA/CA and SOTDMA to improve the performance of MANET. In: IEEE 17th International Conference on Communication Technology (ICCT), pp: 793–799 (2017). https://ieeexplore.ieee.org/abstract/document/8359746
9. Govindasamy, J., Samundiswary, P.: A comparative study of reactive, proactive and hybrid routing protocol in wireless sensor network under wormhole attack. J. Electr. Syst. Inf. Technol. (2017) https://www.researchgate.net/publication/316497034_A_Comparative_Study_of_Reactive_Proactive_and_Hybrid_Routing_Protocol_in_Wireless_Sensor_Network_under_Wormhole_Attack
10. Hashem, I.A., Chang, V., Anuar, N.B., Adewole, K., Yaqoob, I., Gani, A., Ahmed, E., Chiroma H.: The role of big data in smart city. Int. J. Inf. Manag. **36**, 748–758 (2016). https://www.sciencedirect.com/science/article/pii/S0268401216302778
11. Hong, L., Luo, M., Wang, R., Lu, P., Lu, W., Lu, L.: Big data in health care: applications and challenges. Data Inf. Manag. **3**, 175–97 (2018). https://content.sciendo.com/view/journals/dim/2/3/article-p175.xml
12. Islam, S.R., Kwak, D., Kabir, M.H., Hossain, M., Kwak, K.S.: The internet of things for health care: a comprehensive survey. IEEE 3:678–708 (2015). https://ieeexplore.ieee.org/abstract/document/7113786
13. Jain, A.K., Tokekar, V.: Mitigating the effects of Black hole attacks on AODV routing protocol in mobile ad hoc networks. In: International Conference on Pervasive Computing (ICPC) pp. 1–6 (2015). https://ieeexplore.ieee.org/abstract/document/7087174.
14. Kaur, J., Singh, G.: Hybrid AODV algorithm for path establishment in MANET using bio inspired techniques. In: IEEE International Conference on Power, Control, Signals and Instrumentation Engineering ICPCSI, pp. 823–829. https://ieeexplore.ieee.org/abstract/document/8391828.
15. Khan, A., Haa, R.U., Saleem, I., Feroz, T., Saleem, O., Babar, M.I.: Energy based analysis of traditional routing in mobile ad-hoc networks. In: International Conference on Computing, Electronics & Communications Engineering (ICCECE)pp. 281–285 (2018). https://ieeexplore.ieee.org/abstract/document/8658755
16. Kumar, P., Lee, H.J.: Security issues in healthcare applications using wireless medical sensor networks: A survey. Sensors **12**, 55–91 (2012). https://www.mdpi.com/1424-8220/12/1/55
17. Latif, R., Abbas, H., Assar, S.: Distributed denial of service (DDoS) attack in cloud-assisted wireless body area networks: a systematic literature review. J. Med. Syst. **138**:128 (2014). https://link.springer.com/article/10.1007/s10916-014-0128-8

18. Liagkou, V., Kavvadas, V., Chronopoulos, S.K., Tafiadis, D., Christofilakis, V., Peppas, K.P.: Attack detection for healthcare monitoring systems using mechanical learning in virtual private networks over optical transport layer architecture. Computation **7**, 24 (2019). https://www.mdpi.com/2079-3197/7/2/24
19. Mashal, I., Alsaryrah, O., Chung, T.Y., Yang, C.Z., Kuo, W.H., Agrawal, D.P.: Choices for interaction with things on Internet and underlying issues. Ad Hoc Netw **28**, 68–90 (2015). https://www.sciencedirect.com/science/article/abs/pii/S1570870514003138
20. Oh, S., Cha, J., Ji, M., Kang, H., Kim, S., Heo, E., Han, J.S., Kang, H., Chae, H., Hwang, H., Yoo, S.: Architecture design of healthcare software-as-a-service platform for cloud-based clinical decision support service. Healthc. Inf. Res. **21**, 102–10 (2015). https://synapse.koreamed.org/search.php?where=aview&id=10.4258/hir.2015.21.2.102&code=1088HIR&vmode=FULL
21. Prabha, R., Krishnaveni, M., Manjula, S.H., Venugopal, K.R., Patnaik, P.: Qos aware trust metric based framework for wireless sensor networks. Elsevier Science BV (2015). https://eprints.iisc.ac.in/52462/
22. Sanzgiri, K., Dahill, B., Levine, B.N., Shields, C., Belding-Royer, E.M.: A secure routing protocol for ad hoc networks. In: 10th IEEE International Conference on Network Protocols 12:78–87 (2002). https://ieeexplore.ieee.org/abstract/document/1181388
23. Sharma, A.K., Trivedi, M.C.: Performance comparison of AODV, ZRP and AODVDR routing protocols in MANET. In: Second International Conference on Computational Intelligence & Communication Technology (CICT) IEEE pp. 231–236 (2016). https://ieeexplore.ieee.org/abstract/document/7546607
24. Sengupta, A., Sengupta, D., Das, A.: Designing an enhanced ZRP algorithm for MANET and simulation using OPNET. In: Third International Conference on Research in Computational Intelligence and Communication Networks (ICRCICN) pp. 153–156 (2017). https://ieeexplore.ieee.org/abstract/document/8234498
25. Sharma, R., Sharma, T., Kalia, A.: A comparative review on routing protocols in MANET. Int. J. Comput. Appl. **133**, 33–38 (2016). https://www.semanticscholar.org/paper/A-Comparative-Review-on-Routing-Protocols-in-MANET-Sharma-Sharma/2dfba258ced2e104ea1eff54825adf50f59cbc69
26. Sridevi, N., Nagarajan, V.: Enhanced secure wireless communication in MANETs. In International Conference on Communication and Signal Processing (ICCSP), IEEE, pp. 2256–2261 (2017).https://toc.proceedings.com/38107webtoc.pdf
27. Walikar, G.A., Biradar, R.C.: A survey on hybrid routing mechanisms in mobile ad hoc networks. J. Netw. Comput. Appl. **77**, 48–63 (2017). https://www.sciencedirect.com/science/article/pii/S1084804516302430
28. Ashu, Mahajan, R., Zafar, S.: Implementing blockchain security to prevent DDoS attacks in networks. Int. J. Secur. Appl. (IJSIA) **13**(4). ISSN:1738-9976(Print);2207-9629(Online), pp. 27–40, NADIA (2019). https://article.nadiapub.com/IJSIA/vol13_no4/3.html#
29. Yadav, Piyush, Rajeev, A.: A multi-homing based framework against denial of service open threat signaling in healthcare environment. Int. J. Control Automat. 1–18 (2018). https://www.researchgate.net/publication/326990352_Protocols_performance_investigation_using_Ad_Hoc_WLAN_for_healthcare_applications
30. Zamini, M., Hasheminejad, S.M.: A comprehensive survey of anomaly detection in banking, wireless sensor networks, social networks, and healthcare. Intell. Dec. Technol. 13:229–70 (2019). https://content.iospress.com/articles/intelligent-decision-technologies/idt170155
31. Sharma, S.K., Sharma, S.: Improvement over AODV considering QoS support in mobile adhoc networks. Int. J. Comput. Netw. Appl. (IJCNA) **4**, 47–61 (2017). https://scholar.google.com/scholar?cluster=9730674834030613071&hl=en&as_sdt=0,5

An Intelligent Internet of Medical Things with Deep Learning Based Automated Breast Cancer Detection and Classification Model

Mahantesh Mathapati, S. Chidambaranathan, Abdul Wahid Nasir, G. Vimalarani, S. Sheeba Rani, and T. Gopalakrishnan

Abstract In recent decades, breast cancer (BC) is a significant cause of high mortality rate among women. The earlier identification of breast cancer helps to increase the survival rate by the use of appropriate medications. At the same time, internet of medical things (IoMT) and digital mammography finds helpful to diagnose breast cancer effectively in the beginning level itself. This paper presents an intelligent IoMT based breast cancer detection and diagnosis using deep learning model. IoMT based image acquisition process takes place to gather the digital mammogram images. The proposed model performs a set of processes namely preprocessing, K-means clustering based segmentation, local binary pattern (LBP) based feature

M. Mathapati
Department of Computer Science and Engineering, Raja Rajeswari College of Engineering, Bengaluru, India
e-mail: manteshkrishna@gmail.com

S. Chidambaranathan
Computer Applications, ST. Xavier's College Autonomous, Palayamkottai, India
e-mail: scharan2009@gmail.com

A. W. Nasir
Department of Electronics and Communication Engineering, CMR Institute of Technology, Bengaluru, India
e-mail: abdulwahid.n@cmrit.ac.in

G. Vimalarani
Department of Electronics and Communication Engineering, Hindustan Institute of Technology and Science, Chennai, India
e-mail: gvrani@hindustanuniv.ac.in

S. Sheeba Rani
Department of Electrical and Electronics Engineering, Sri Krishna College of Engineering and Technology, Coimbatore, India
e-mail: sheebaranis@skcet.ac.in

T. Gopalakrishnan (✉)
School of Computer Science and Engineering, Vellore Institute of Technology, Vellore, India
e-mail: gopalakrishnan.ct@gmail.com

A. E. Hassanien et al. (eds.), *Cognitive Internet of Medical Things for Smart Healthcare*, Studies in Systems, Decision and Control 311,
https://doi.org/10.1007/978-3-030-55833-8_11

extraction and deep neural network (DNN) based classification. The presented LBP-DNN model has the capability of effectively detecting and classifying breast cancer from mammogram images. The LBP-DNN model has been validated using MIAS database and an extensive comparative analysis is carried out to evaluate its performance. The experimental results ensured the superior performance of the LBP-DNN model with the maximum sensitivity of 71.64%, specificity of 75.87% and accuracy of 70.53%.

Keywords Digital mammogram · Breast cancer · Deep learning · Feature extraction · Internet of medical things · MIAS database

1 Introduction

Internet of Medical Things (IoMT) is an incorporation of medicinal gadgets and applications which links the healthcare information technology systems by the use of networking technologies. It reduces the unwanted hospital visits and load on healthcare systems by linking patients to their doctors and enables the communication of medicinal data over the secure network. At recent times, Breast Cancer (BC) is the most common cancer diagnosed in women globally and leads to 23% of cancer patients and 14% of deaths amongst women based on cancer [1]. On comparing the diagnosis of symptoms, mammogram helps doctors exactly analyse the BC. The goal of Mammography screening program is to minimize the BC death rate by diagnosing the disease at the starting stage. At once, BC is recognized in the starting stage and it is better to begins the prescription and reduce the mortality of the patient. The radiologists observe the mammogram images during screening technique. A skilled radiologist is mandatory to formulate flawless analysis, but sometimes victims may be spotted normal cases by mistake. Humans tend to do flaws and the corrupted conclusion could end up the cases into hopeless stage of BC. Computer Aided Diagnosis (CAD) structure is supportive devices to the physicians to minimize the corrupted diagnosis percent and accuracy is enhanced. CAD structure contains a variety of processing approaches namely classification, feature extraction, segmentation and pre-processing. All processing stages are similar important in diagnosis.

The most essential goal is to create an entire computerized classifier approach for labelling the mammogram images that aid to detect the disease in CAD system. Deep CNN (DCNN) gives Deep Learning (DL) architecture which is utilized for explaining classifiers issues in computer vision applications. DCNN resolves the classifiers issues by removing the characters through different convolution filters. The presentation of the off-shelf classification techniques namely, Support Vector Machine (SVM), Random Forest (RF) are highly subjected by the class of training data and glance at the issues by considering the feature. The process of testing and selecting certain number of features are considered to be tedious operation.

DL technique defeats this issue by acquiring better information on the application field by DL with the enormous range of database. Guiding DCNN from scratch is complex always due to gradient diminishing as well as convergence issues [2]. Though the information of DCNNs is monotonous and time consuming, they have capability of earning high useful descriptions without individual's involvement. In DCNN, feature extraction and classification are enhanced to achieve better performance. Automatic classification of mammographic lumps are tedious because of the existence of homogeneous images. Nowadays several research techniques are applied in the BT Diagnosis domain by utilizing hand crafted figures features and conventional machine learning classification techniques. Zhang et al. [3] planned a spatially invariant NN for classifying Region of Interest's (ROI's) as normal and abnormal tissues.

Wu et al. [4] have established a shallow Back Propagation NN (BPNN) for classifying benign and malignant tumors according to the predominant features monitored by the professional radiologists. Oliver et al. [5] executed a knowledge-based classification with the help of Adaboost SVM for the automatic mammographic image classification. While the CNN is not essential for handcrafted features,it acquires the feature from the image and divide the images with the help of trained CNN method. DCNN's is composed of benchmark details over massive fine-tuned object classification operations.

Hafemann et al. [6] assured that the DCNN technique is capable of recognizing the forest species datasets according to the textures, where the CNN model is classified into 2 forest species with evident results. As the cancer patients have maximum lumps in breast region, morphological dependent features such as texture and density is used for examining the anomalies of mammogram images. In recent times, transfer-learning method was applied in fine-tuning of pre-trained DCNN approaches to classify mammography lesions. The fine-tuned DCNN method has been applied in CAD system and provided benchmark outcomes [7]. The developers ensured that the mammography classification task on the basis of representation learning offers enhanced computation when related with feature relied classification system. Likewise, the hybrid classification approach has provided optimal classification results under the application of learned as well as hand engineered depictions [8]. Many developers have represented the efficiency of CNN dependent CAD system has been enhanced by integrating intensity-based as well as learned features.

Sapate and Talbar [9] concentrates on the impacts in resolving the pectoral muscle extraction problem in pre-processing task to predict the BC at primary stage using MGMs. On the other hand, a rare case with processing complexity is computed with maximum dimension. Sivakumar et al. [10] applied an improvised artificial bee colony (ABC) optimization method for detecting the borders and to identify the nipple spot. The newly developed EABCO technique was employed for resolving the extraction of malicious region from MGM image under the application of bilateral subtraction. The simulation outcome as well as comparison depends upon the bilateral model that produce optimal outcome.

Maitra et al. [11] employed massive count of algorithms like binary homogeneity improvement, reputed convolution models, image segmentation, seeded region development, pectoral muscle prediction algorithm, and breast boundary prediction method. The integration of models has been offered magnificent prediction of mass in MGMs images such as cyst, calcification, or fatty masses. Nithya et al. [12] projected region developing model with top-down technique. It divides the ROI from the background image. From the classified images, Genetic Algorithm (GA) is applied. At this point, the mammography is referred as better ways to identify BC in the initial stage. Improvement of mammography image is processed by median filter. The entire segmentation is managed by region developing method.

Percha et al. [13] provides models like plasmon coupling effects, BPNN, as well as prediction approaches. It is applied with micro-calcification technique, CAD algorithm, and clustering method for detecting the micro-calcification clusters and offers the maximum final outcome. Rubin et al. [14] defines the way of differentiating the speculated masses and cyst. The prominent anomalies which indicate BC are masses and calcifications. Gray level co occurrence matrices (GLCM), a texture feature has been applied for classifying masses and healthy breast tissues from input image. Angayarkanni et al. [15] exploits edge prediction and segmentation technique for detecting tumor in the breast from mammogram images.

This paper devises a new breast cancer detection and diagnosis using DL model. The proposed model involves a series of processes such as IoMT based image acquisition, preprocessing, K-means clustering based segmentation, local binary pattern (LBP) based feature extraction, and DNN based classification. The proposed LBP-DNN model effectively detects and classifies the breast cancer from mammogram images. The LBP-DNN model has been validated using MIAS database and a widespread comparative examination takes place for assessing the classifier outcome.

2 The Proposed LBP-DNN Model

The working principle of the proposed LBP-DNN model is depicted in Fig. 1. As shown, the IoMT based image acquisition process takes place to gather the digital mammogram images. Then, the gathered images undergo preprocessing for the purpose of eliminating artifacts exists in the image and improve the image quality to a certain extent. Followed by, K-means clustering technique is applied for the segmentation of the images. Next to that, the LBP technique is executed for feature extraction in the segmented image. At the end, the DNN based classification process is applied to diagnose and classify the mammogram image.

2.1 K-means Clustering Based Segmentation

Once the input image is gathered and pre-processed, segmentation process takes place to detect the affected regions exist in the mammogram image with the help of K-means clustering model. Some of the procedures involved in this method is provided in the following [16]:

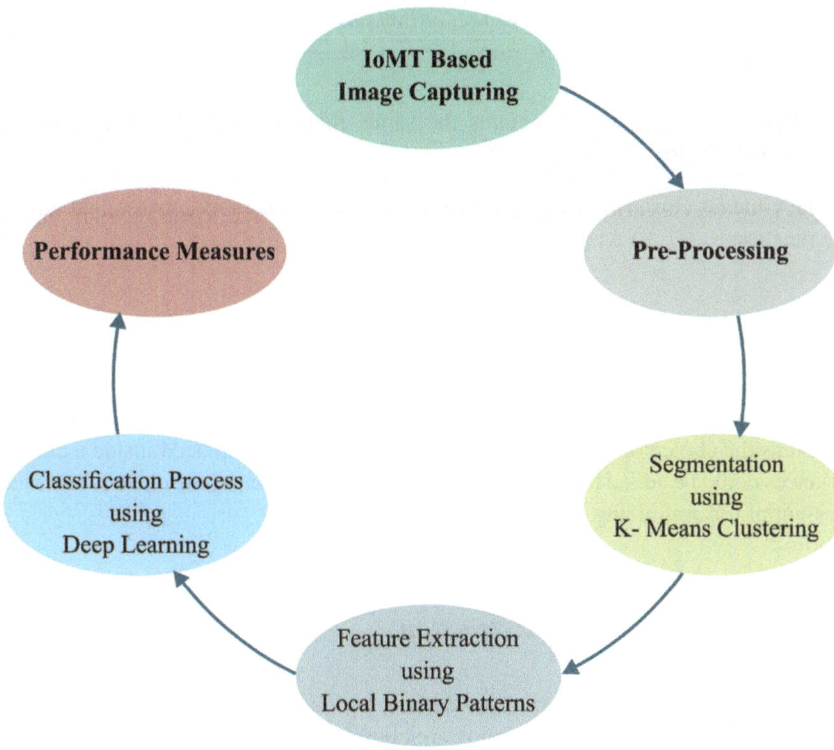

Fig. 1 Overall process of proposed method

1. Choose the count of clusters J < with primary cluster centroid $v_i, i = 1, 2, \ldots k$.
2. Divide the input data points into k clusters by allocating every data point x_j to nearby cluster centroid v_i using elected distance value, where Euclidean distance is described as,

$$d_{ij} = \left|\left|x_j - v_i\right|\right| \tag{1}$$

where $X = \{x_1, x_2, \ldots x_n\}$ shows input dataset.
3. Calculate a cluster assignment matrix U which represents the division of data points along with binary membership value of jth data point to ith cluster in which $U = \left|u_{ij}\right|$, where u_{ij} in {0, 1 } for all i, j

$$\sum_{i=1}^{k} u_{ij} = 1\; for\; all\; j\; and\, 0 < \sum_{j=1}^{n} u_{ij} < n\;\; for\; all\, i. \tag{2}$$

4. Re-estimate the centroid with the help of membership values by

$$n_i = \frac{\sum_{j=1}^{n} u_{ij} x_j}{\sum_{ij=1}^{n} u_{ij}} \quad for\ all\ i \tag{3}$$

5. When a cluster centroid remains the same, then the existing iteration is not terminated; else, return to step 2.

The k-means clustering model generalizes the sum-of-squared-error relied objective function $J_w(U, v)$ which is expressed as,

$$J_w(U, v) = \sum_{i=1}^{h} \sum_{j=1}^{n} ||x_j - v_i||^2 \tag{4}$$

The k-means clustering model is highly sensitive to the primary cluster assignment as well as the choice of distance measure. The extra criterion placed inside a cluster variances could be added to the objective function for exploiting the clusters k, which is essential for optimizing the objective function.

2.2 LBP Based Feature Extraction

LBP model is applied in diverse applications and is employed for medical image diagnosis [17]. In LBP, the histograms are combined into an individual vector and the vector is assumed as a pattern vector. In alternate applications, the integration among LBP texture features as well as Self-Organizing Map is used for identifying the model's efficiency. LBP is defined as operator for texture definition which depends upon the signs of variations among neighbour pixels as well as central pixels. For every pixel value present in the image, a binary code has been attained with the help of thresholding the neighbourhood using middle pixel. The binary code is referred as binary pattern. Hence, neighbour pixel is 1 when the pixel value is higher when compared to threshold value, and it is 0 when the pixel value is minimum than threshold. Besides, the histogram would be developed for computing the frequency values of binary patterns. Every pattern implies the viability of binary pattern that exists in an image. The count of histogram bins is based on the number of pixels engaged in LBP determination. When LBP applies 8 pixels, the number of histogram bin becomes 2^8 or 256.

The fundamental method of LBP operator applies the measure of centre pixel as a threshold to the 3×3 neighbour pixels. Threshold operation could develop a binary pattern that shows texture features. The function of LBP can be expressed as,

$$LBP(u_c, v_c) = \sum_{n=0}^{7} 2^n g(I_n - I(u_c, v_c)) \tag{5}$$

$LBP(u_c, v_c)$ defines the LBP value at intermediate pixel (u_c, v_c). I_n and $I(u_c, v_c)$ are values of adjacent pixel as well as centre pixel correspondingly. Index n refers the index of neighbour pixels. The function $g(u)$ would be 0 when $u < 0$ and $g(u) = 1$ if $u \geq 0$. The adjacent pixels would be 0, when the values are minimum when compared with threshold. In contrast, it will be 1, when the neighbour pixels are higher than threshold. The LBP value is determined by using scalar multiplication among the binary and weight matrices. Then, consolidation of multiplication results has been applied to imply the LBP value.

2.3 DNN Based Classification

Artificial Neural Network (ANN) mechanism is applied diverse layers of hidden units and outputs termed as DL. It is comprised of 2 phases such as,

- Pre-training
- Fine-tuning

2.3.1 Pre-training Stage

It is said to be the deep structure and significant FFNN, Deep Belief Network (DBN) were applied for generating input layer to output layer by higher number of hidden layers with additional layers. According to the application of DBN, the hidden units distinguish the network and it generates activation functions [18]. Also, RBM is deployed to solve the problems of feasible activation function production.

2.3.2 Restricted Boltzmann Machine

RBM belongs to Markov random domain that is comprised of single layer of stochastic hidden units such as single layer of stochastic visible units. Initially, upload the clear units for training the RBM vector as defined in Eq. (6):

$$F(v, h) = -\sum_{k=1}^{K}\sum_{l=1}^{L} W_{kl} v_k h_l - \sum_{k=1}^{K} \alpha k v k - \sum_{l=1}^{L} \beta_l h_l \tag{6}$$

The extension of Eq. (8) is given in the following; W_{kl} is the symmetric interaction between the visible unit vk and the hidden unit h_l, α, β are bias terms, K, L, means the number of clear and hidden units. The subordinate of the log likelihood of a preparation vector which considers a weight is inconsistent. From the hidden units of RBM, the immediate response is not required for acquiring simple impartial sample from $(V_k, h_l)_{data}$

$$\rho(h_{l=1}|v) = \delta\left(\sum_{k=1}^{K} W_{kl} v_k + \alpha_l\right) \tag{7}$$

where $\delta(x)$ means the logistic sigmoid function. $\frac{1}{(1+\exp(x))}$, v_k, h_l defines unbiased sample.

2.3.3 Updating Process

The hidden unit is extended while the visible units are considered to be concurrent from given visible and hidden units. It results in difficult steps as given below.

$$\triangle\ W_{kl}\theta(v_k h_l)_{data} - (v_k h_l)_{reconstruction} \tag{8}$$

Here, RBM is provided with training. A distinct RBM is stacked over a frame with the help of multilayer model. The input visible layer is ordered as a vector, the qualities for units that are effectively developed with RBM layers that are distributed using distributed methods in recent weights and biases. The attained DNN weights are satisfied with fine-tuning phase.

2.3.4 Fine Tuning Phase

It is carried out on the basis of BP model. The clinical images can be segregated using 2 phases; output layer has been arranged from top of DNN. In addition, N number of input neurons and 3 hidden layers were employed in DL approach. The improvised weight has to be attained at the training phase; also, the training data set is used for initializing BP using the weights acquired from pre-training level. At last, Eq. (8) can be applied with minimum error value was determined and higher accuracy of DL classification can be accomplished using best weight.

3 Experimental Evaluation

The experimental outcome of the LBP-DNN model undergoes validation against MIAS database. Figure 2 shows a sample set of test images from the applied database.

The Mammographic Image Analysis Society (MIAS) is an industry of UK research groups interested in the learning mammograms and has produced a database of digital mammograms [19]. Films captured from the UK National Breast Screening Programme have been digitised to 50-μm pixel edge using a Joyce-Loebl scanning microdensitometer, a device linear in the optical density ranks from 0 to 3.2 and implying every pixel with an 8-bit word. The database is comprised of 322 digitised films and accessible on 2.3 GB 8 mm (ExaByte) tape. Also, it contains radiologist's

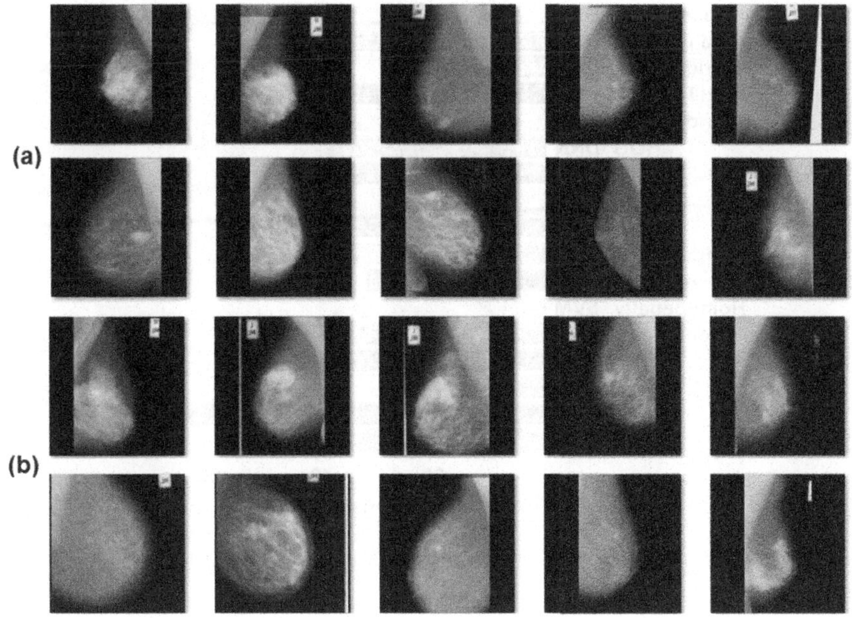

Fig. 2 Sample images. **a** Benign. **b** Malignant

"truth"-markings on the locations of anomalies which might be represented. The database is limited to a 200-μm pixel edge and padded/clipped where images are 1024 × 1024. Mammographic images are accessible through the Pilot European Image Processing Archive (PEIPA) at the University of Essex.

Figures 3, 4 and 5 made a detailed comparative study of the classiifer results of the LBP-DNN with existing models. Figure 3 shows the sensitivity analysis of the proposed LBP-DNN model with the existing techniques on the applied dataset. It is apparent that the Energy-NB model has offered a minimum sensitivity of 26.4%. besides, the hybrid feature-NB model has attained slightly higher sensitivity of 28%. At the same time, the Homogeneity-NB model has leads to somewhat higher sensitivity of 32%. Along with that, the Homogeneity-SVM has attained even more sensitivity of 37.6%. In line with, the HIG-NB, HIG-NN and HOG-DNN models have attained low and closer sensitivity values of 38%, 38.4% and 38.4% respectively. Simultaneously, the Homogeneity-NN and Homogeneity-DNN models have reached to moderate and equivalent sensitivity value of 54.4%.

In the same way, the Energy-NN and Energy-DNN models have tried to exhibit manageable outcome with the identical sensitivity of 56.8%. Also, the HOG-SVM model has produced a slightly acceptable sensitivity of 57.6%. Moreover, the Hybrid Features-SVM, Hybrid Features-NN and Hybrid Features-DNN models have resulted to a somewhat higher and closer sensitivity values of 58%, 58.4% and 58.4% respectively. Furthermore, the Energy-SVM model has reached to a near optiaml

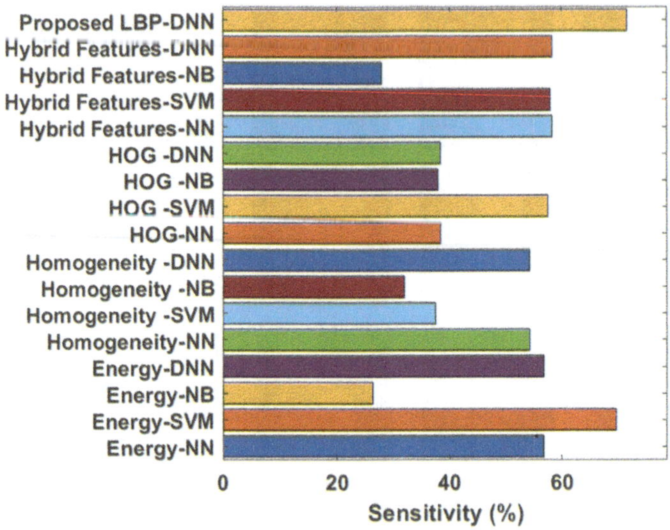

Fig. 3 Sensitivity analysis of proposed LBP-DNN method with other models

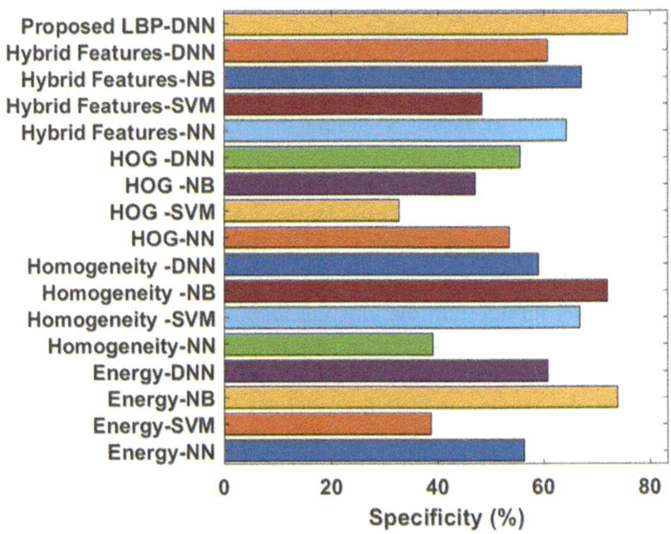

Fig. 4 Specificity analysis of proposed LBP-DNN method with other models

sensitivity value of 69.6%. However, the LBP-DNN model has showcased better results over the compared methods with the maximum sensitivity value of 71.64%.

Figure 4 implies the specificity analysis of proposed LBP-DNN model with the previous models on the provided dataset. It is evident that the HOG-SVM method has provided a lower specificity of 32.8%. On the other hand, the Energy-SVM approach

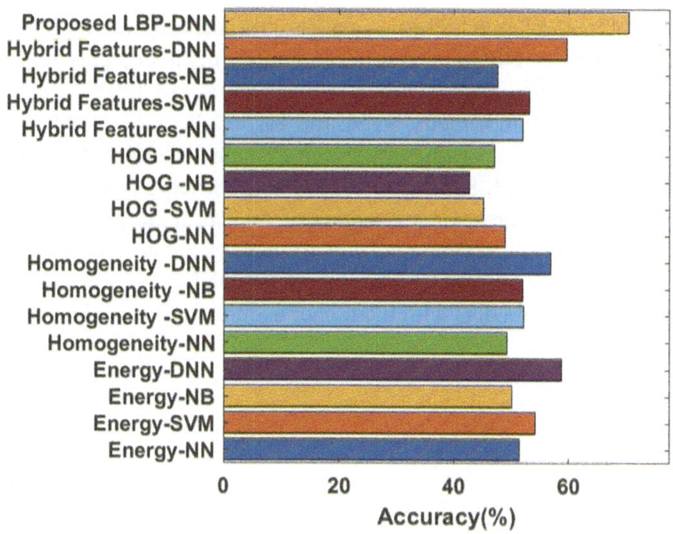

Fig. 5 Accuracy analysis of proposed LBP-DNN method with other models

has accomplished moderate specificity of 38.8%. Simultaneously, the Homogeneity-NN technique provides considerable specificity of 39.2%. In line with this, the HOG-NB has achived slightly better specificity of 47.2%. On the same way, the Hybrid Features-SVM, HOG-NN and HOG-DNN models have reached minimum and closer specificity values of 48.40%, 53.6% and 55.6% correspondingly. At the same time, the Energy-NN and Homogeneity-DNN methodologies have attained a gradual specificity value of 56.4% and 59.2% respectively. Similarly, the Hybrid Features-DNN and Energy-DNN approaches have attempted to show manageable outcome with the similar specificity of 60.8%. In addition, the Hybrid Features-NN model has generate suitable specificity of 64.4%. Furthermore, the Homogeneity-SVM, Hybrid Features-NB and Homogeneity-NB models have provided reasoanble and near specificity values of 66.8, 67.2 and 72% correspondingly. Moreover, the Energy-NB model has attained a near optiaml specificity value of 74%. Hence, the LBP-DNN model has exhibited acceptable results than the compared methods with the higher specificity value of 75.87%.

Figure 5 demonstrates the accuracy analysis of the presented LBP-DNN model with the traditional models on the applied dataset. It is pointed that the HOG-NB method has providded a least accuracy of 42.6%. Followed by, the HOG-SVM approach has accomplished gradual accuracy of 45.2%. Meanwhile, the HOG-DNN technology tends to provide maximum accuracy of 47%. In line with this, the Hybrid Features-NB has achieved even better accuracy of 47.6%. Along with that, the HOG-NN, Homogeneity-NN and Energy-NB models have obtained low and near accuracy values of 49%, 49.2% and 50.2% respectively. At the same time, the Energy-NN model has achieved a considerable accuracy value of 51.4%. LIke-wise, the Homogeneity-NB and Hybrid Features-NN methodologies have attempted to showcase reasonable outcome with the same accuracy of 52%.

Additionally, the Homogeneity-SVM and Hybrid Features-SVM models have generated a managebale accuracy values of 52.2 and 53.2% correspondingly. In addition, the Energy-SVM, Homogeneity-DNN and Energy-DNN techniques have attained better and closer accuracy values of 54.2%, 56.8% and 58.8% respectively. Moreover, the Hybrid Features-DNN framework has accomplished to a closer optimal accuracy value of 59.6%. Thus, the LBP-DNN model has depicted good results than the related models with the higher accuracy value of 70.53%.

4 Conclusion

This paper has devised a new breast cancer detection and diagnosis using DL model. The proposed model involves a series of processes such as IoMT based image acquisition, preprocessing, K-means clustering based segmentation, LBP based feature extraction, and DNN based classification. Initially, IoMT based image acquisition process takes place to gather the digital mammogram images. Then, the gathered images undergo preprocessing to remove the noise exist in the image and improve the image quality to a certain extent. Followed by, K-means clustering technique is applied for the segmentation of the images. Next to that, the LBP technique is executed to extract the features exist in the segmented image. At the end, the DNN based classification process is applied to diagnose and classify the mammogram image. The presented LBP-DNN model has the capability of effectively detecting and classifying breast cancer from mammogram images. The LBP-DNN model has been validated using MIAS database and an extensive comparative analysis is carried out to evaluate its performance. The experimental results ensured the superior performance of the LBP-DNN model with the maximum sensitivity of 71.64%, specificity of 75.87% and accuracy of 70.53%.

References

1. Whang, J.S., Baker, S.R., Patel, R., Luk, L., Castro, A., III.: The causes of medical malpractice suits against radiologists in the United States. Radiology **266**(2), 548–554 (2013)
2. Glorot, X., Bengio, Y.: Understanding the difficulty of training deep feedforward neural networks. In: Proceedings of the Thirteenth International Conference on Artificial Intelligence and Statistics, pp. 249–256 (2010)
3. Zhang, W., Doi, K., Giger, M.L., Wu, Y., Nishikawa, R.M., Schmidt, R.A.: Computerized detection of clustered microcalcifications in digital mammograms using a shift-invariant artificial neural network. Med. Phys. **21**(4), 517–524 (1994)
4. Wu, Y., Giger, M.L., Doi, K., Vyborny, C.J., Schmidt, R.A., Metz, C.E.: Artificial neural networks in mammography: application to decision making in the diagnosis of breast cancer. Radiology **187**(1), 81–87 (1993)

5. Oliver, A., Torrent, A., Lladó, X., Tortajada, M., Tortajada, L., Sentís, M., Freixenet, J., Zwiggelaar, R.: Automatic microcalcification and cluster detection for digital and digitised mammograms. Knowl.-Based Syst. **28**, 68–75 (2012)
6. Hafemann, L. G., Oliveira, L.S., Cavalin, P.: Forest species recognition using deep convolutional neural networks. In: 2014 22nd International Conference on Pattern Recognition (ICPR), pp. 1103–1107 (2014)
7. Chougrad, H., Zouaki, H., Alheyane, O.: Deep convolutional neural networks for breast cancer screening. Comput. Methods Prog. Biomed. **157**, 19–30 (2018)
8. Arevalo, J., González, F.A., Ramos-Pollán, R., Oliveira, J.L., Lopez, M.A.G.: Representation learning for mammography mass lesion classification with convolutional neural networks. Comput. Methods Prog. Biomed. **127**, 248–257 (2016)
9. Sapate, S., Talbar, S., et al.: Pectoral muscle extraction algorithms applied to digital mammograms. Inst. Politec. Nacional Centro de Innovation y Desarrollo Technology en Comput. Mexico **7**(2)3, 5–6 (2005)
10. Sivakumar, R., Karnan, M.: Diagnose breast cancer through mammograms using EABCO algorithm. Int. J. Eng. Technol. **4**, 302–307 (2012)
11. Maitra, I.K., Nag, S., Bandyopadhyay, S.K.: Automated digital mammogram segmentation for detection of abnormal masses using binary homogeneity enhancement algorithm. Indian J. Comput. Sci. Eng. **2**(3), 416–427 (2011a)
12. Nithya, R., et al.: Improving performance of breast cancer detection for mammography image. Int. J. Adv. Comput. Sci. Appl. **25**(5) (2011)
13. Percha, B., et al.: Computer aided detection algorithm for digital mammogram images. Int. J. Comput. Trends Technol. (IJCTT) **2**(2) (2012)
14. Maitra, I.K., Nag, S., Bandyopadhyay, S.K.: Identification of abnormal masses in digital mammography images. Int. J. Comput. Graph. **2**(1), 17–30 (2011b)
15. Angayarkanni, N., Kumar, D., Arunachalam, G.: The application of image processing techniques for detection and classification of cancerous tissue in digital mammograms. J. Pharm. Sci. Res. **8**(10), 1179 (2016)
16. Rajini, N.H., Bhavani, R.: Computer aided detection of ischemic stroke using segmentation and texture features. Measurement **46**(6), 1865–1874 (2013)
17. Prakasa, E.: Texture feature extraction by using local binary pattern. INKOM J. **9**(2), 45–48 (2016)
18. Raj, R.J.S., Shobana, S.J., Pustokhina, I.V., Pustokhin, D.A., Gupta, D., Shankar, K.: Optimal feature selection-based medical image classification using deep learning model in internet of medical things. IEEE Access **8**, 58006–58017 (2020)
19. https://www.mammoimage.org/databases/

Internet of Medical Things (IoMT) Enabled Skin Lesion Detection and Classification Using Optimal Segmentation and Restricted Boltzmann Machines

A. Peter Soosai Anandaraj, V. Gomathy, A. Amali Angel Punitha, D. Abitha Kumari, S. Sheeba Rani, and S. Sureshkumar

Abstract In recent times, Internet of Medical Things (IoMT) and cloud enabled healthcare applications and services finds helpful for effective decision-making. Melanoma is the serious kind of skin cancer, results to high death rate. Earlier identification of skin cancer can leads to maximum survival rate. But the diagnosis process becomes difficult and expensive because of the need of medical experts and complex medical equipments. To overcome this issue, the latest developments in IoMT based decision making system with maximum performance can be used. This study introduces a new IoMT based skin lesion detection and classification model using Optimal Segmentation and Restricted Boltzmann Machines (RBM), named OS-RBM model. The proposed OS-RBM model involves a series of steps namely image acquisition, gaussian filtering (GF) based preprocessing, segmentation, feature extraction, and classification. Then, optimal segmentation using artificial bee colony (ABC) with

A. Peter Soosai Anandaraj
Ganapathy Chettiar College of Engineering and Technology, Paramakudi, India
e-mail: anandsiriya@gmail.com

V. Gomathy · S. Sheeba Rani (✉)
Department of EEE, Sri Krishna College of Engineering and Technology, Coimbatore, India
e-mail: sheebaranis@skcet.ac.in

V. Gomathy
e-mail: gomathyv@skcet.ac.in

A. Amali Angel Punitha
Ultra College of Engineering and Technology, Madurai, India
e-mail: kirubajose@gmail.com

D. Abitha Kumari
Department of Computer Science and Engineering, Sethu Institute of Technology, Virudhunagar, India
e-mail: abithajustin2005@gmail.com

S. Sureshkumar
Department of CSE, Sri Krishna College of Engineering and Technology, Coimbatore, India
e-mail: ssureshkumar@skcet.ac.in

A. E. Hassanien et al. (eds.), *Cognitive Internet of Medical Things for Smart Healthcare*, Studies in Systems, Decision and Control 311,
https://doi.org/10.1007/978-3-030-55833-8_12

kapur's thresholding takes place. Besides, histogram and texture feature extraction will be carried out. Finally, RBM is applied as a classifier to detect and classify the existence of skin lesion in the dermoscopic images. A detailed simulation analysis takes place for ensuring the better outcome of the OS-RBM model and the results are assessed under diverse performance measures. The experimental outcome ensured the effective classification performance of the OS-RBM model with the maximum sensitivity of 96.43%, specificity of 97.95% and accuracy of 95.68%.

Keywords Skin lesion · Segmentation · Deep learning · Classification · Restricted Boltzmann machine

1 Introduction

Cognitive computing is computational technology which offers an artificial physical response, authorizing a subject to test events and numerous activities similar to those that can be recognized in real time. Internet of Medical Things (IoMT) offers different ways to provide quality treatment and enhance the public health [1]. The IoMT comprises a set of medical gadgets which observes the patient condition and perform decision making. Skin Cancer (SC) occurs because of abandoned unusual growth in skin cells, which has the possibilities of affecting other parts in body. SC is the most familiar and fast expanding types of cancers [2]. In general, it is categorized into 2 modules namely the Malignant-Melanoma and Benign. The melanoma is the harmful cancer type and it has a great morbidity and mortality worldwide [3]. Although it accounts for only 1% of all types of cancers' cases, it has the highest death rate. In general, the skin cancer is categorized into 2 classes namely Malignant-Melanoma and Benign. Among these 2 modules, melanoma is defined as fatal type of cancer which results in extreme deadly rate. Though it reports only one percentage of the entire category of cancers cases, it has the maximum mortality.

American Cancer Society (ACS) predicted that 87,110 novel patients of melanoma would analyse and 9,730 humans might die in America in 2017. Likewise, 132,000 melanoma and 3 million non-melanoma patients were accounted globally [4]. Despite, melanoma is fixable cancer in common, if identified at premature stage. The 5-year survival ratio is about ninety-six percentage if noticed at stage-I, while the ratio is reduced to only five percentages when it is sensed at stage-IV [5]. Because of rising ratio, it got big awareness from both the health care practitioner and researcher. Hence, the vital aim of a dermatologist is to analyse the melanoma at its premature level. Expert dermatologist analyse perfectly though it is hard to characteristics difference between of melanocytic and non-melanocytic skin lesions. In medical analysis, dermatologist utilizes dermoscopy for detection of melanoma. Dermoscopy is a non-invasive image diagnosis method for learning of pigmented skin lesions in dermatology [6].

At first, Dermatologist applies the gel on skin lesion part and make use of a magnifier that increases the skin lesion a number of times exhibiting the pigmented network more obvious than by the human eye. Further it employed to identify the melanoma with different diagnosing algorithms namely Menzies, ABCD rule [7] and 7-point checklist techniques. On the other hand, it is strongly dependent on the individual visualization, formal guidance, and medical practice of the dermatologist. Therefore, dermoscopy has about seventy-five percentage accuracy in melanoma diagnose and to enhance the victory speed in melanoma diagnose, programmed dermoscopic image analysis system could employ. Computer-aided skin lesion detection and classifiers have numerous approaches like pre-processing, classification, feature extraction, and segmentation. The most important challenge in the computer-aided detection (CAD) system is lesion part segmentation and frequent features selection from the unprocessed dataset. Furthermore, these systems apply unique imaging modalities, and every one holds its own restrictions which make this work more difficult.

Recently, the studies carried out were based on the disease analyzing models which are classified into manual, automatic, and hybrid approaches. The primary technique needs visual observation of patient's skin using a well-equipped devices and physicians [8]. For computational technologies, lesion classification is carried out by applying a system, and limits the subjective solutions. The combinations of manual and automated models are named as hybrid approaches. As the skin type and image quality are different, then image classification is processed with the help of automated technique which is highly difficult. Various pre-processing task has been applied for image classification.

The feature extraction is processed with the help of attributes and, thus image classification is carried out. Melanocytic lesions were categorized into 3 classes namely, common nevi, atypical nevi, and melanomas. Initially, common nevi are composed of symmetrical shapes, with definite edges, uniform color such as brown or skin color [9]. Secondly, atypical nevi are dissimilar with uneven borders and difference in shape and color. The diameter is maximum than the width of previous model. Melanoma is same as atypical nevi. Typically, melanoma is initiated as superficial, indolent tumors. At this point, tumors are developed rapidly towards the depth of the dermis. On contrary, the second phase has maximum possibility of metastasis, when diagnosed from horizontal phase, which can be cured by processing surgical task [10].

The atypical nevi are composed of pathogenic association with melanoma, a diagnosis consistent with real-time patient's health condition and helps in making better solutions. Some of the mobile connected dermatoscopic tools in medical sector are DermLite and HandyScope. Hence, it is not applicable due to the higher expense and appropriate training has to be provided. There are few studies about IoT system which eliminates the subjectivity in the classification of skin lesions. Abuzaghleh et al. [11] has projected a real-time alert mechanism which is applied for recovering skin burns and automatic image analysis module. Thus, the process requires pre-processing steps as well as lesion segmentation to eliminate noise in the image.

Do et al. [12] deployed image classification of skin lesions captured by mobile phones with worst lighting as well as focal conditions. The work substantiated with the help of images from 184 images. Sahu et al. [13] projected a portable guidance with the application of a Raspberry Pi for skin lesion classification. Though the network connectivity is not essential, the model is minimized with battery duration. As mentioned above, the pre-processing step is highly essential. The study with considerable results which does not apply the pre-processing steps as introduced by Hosny et al. [14]. Hence, it does not apply the IoT system for diagnosing the disease.

This study introduces a new IoMT based skin lesion detection and classification model using Optimal Segmentation and RBM, named OS-RBM model. The proposed OS-RBM model collects the dermoscopic images using IoMT devices. Then, gaussian filtering (GF) based preprocessing is done to remove the unwanted details exist in the image. Followed by, optimal segmentation using ABC with kapur's thresholding takes place. Besides, histogram and texture feature extraction will be carried out. At the end, RBM is applied as a classifier to detect and classify the existence of skin lesion in the dermoscopic images. A detailed simulation analysis takes place using benchmark dermoscopic images and the results are assessed under diverse performance measures.

2 The Proposed OS-RBM Model

Figure 1 demonstrates the flow diagram of the provided OS-RBM model. The input dermoscopic images are initially captured by the IoMT devices. Then, GF is applied to eliminate the unwanted details and enhance the image quality. Then, ABC-KR based OS technique is executed to segment the pre-processed images. Simultaneously, the histogram and texture features exist in the segmented image are extracted and finally classified by RBM model.

2.1 *Preprocessing*

GF or gaussian blur is a linear filter utilized for smoothing the image that usually decreases image noise. For computing the weighted average, it utilizes Gaussian distribution, a probability based function along with bell-function as its shape. It is implementing in the case of both 1-dimensional and 2-dimensional cases. The cut-off frequency of the filter is to get the ratio among the rate of sample f_s and the standard deviation 0. It is related as following.

$$f_c = \frac{f_s}{\sigma} \tag{1}$$

The 1D GF are provided with the equation:

$$g(x) = \frac{1}{\sqrt{2\Pi}\sigma} e^{\frac{-x^2}{2\sigma^2}} \tag{2}$$

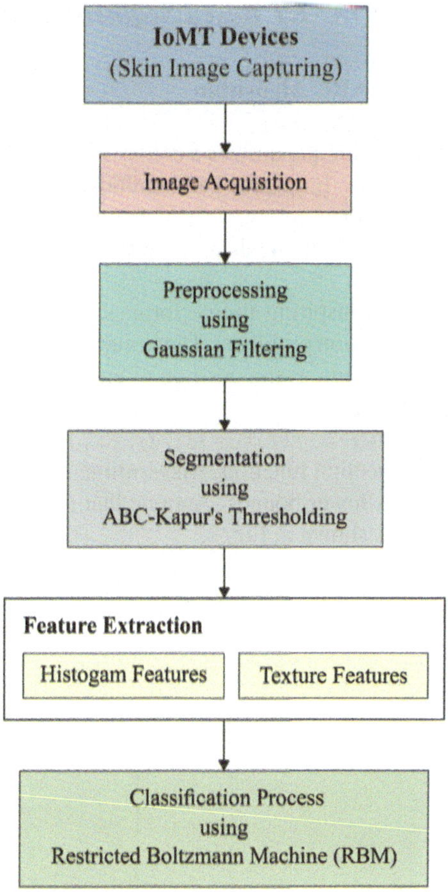

Fig. 1 Overall process of proposed OS-RBM method

The impulse response of the 1D GF is provided with:

$$g(x) = \frac{1}{\sqrt{2\Pi}\sigma} e^{\frac{-\sigma^2 u^2}{2}} \tag{3}$$

If distribution is trade in 2D images, it makes a surface contours as concentric circles containing Gaussian distribution from centre point. The distributed values are utilized for preparing the convolution matrix that are implementing to original image. A weighted average of all pixel neighbourhoods is obtained and put as novel value of the pixel. The original value of the pixels obtains the maximum weight and distance of nearest pixel increases, it is receiving minimum weights.

2.2 ABC-KT Based Segmentation

2.2.1 ABC Algorithm

ABC technique utilizes a colony of artificial bees. These bees are categorized into 3 categories: Employee bees, Onlookers and Scouts. Every employee bee is connected through the food source, which is comparatively equal to the food location. Onlooker bees distribute the data of the food sources initiated by employee bees to select the best ones.

With inspiring the performances of bee swarm, ABC technique is applied for optimization problems. The nectar number of a food source is equivalent to the value of the result, as the position of employee bees or onlookers implies the solution, as given below.

$\min f = f(x)x = (x_1, x_2, \wedge x_m) \in S$ here, $S = [x_{iL}, x_{iH}]$. Where, f indicates the objection function, x is m -dimensional variable, $[x_{iL}, x_{iH}]$ denotes the upper as well as lower bounds. Assume that the amount of the employee bees and onlooker bees as shown in Fig. 2.

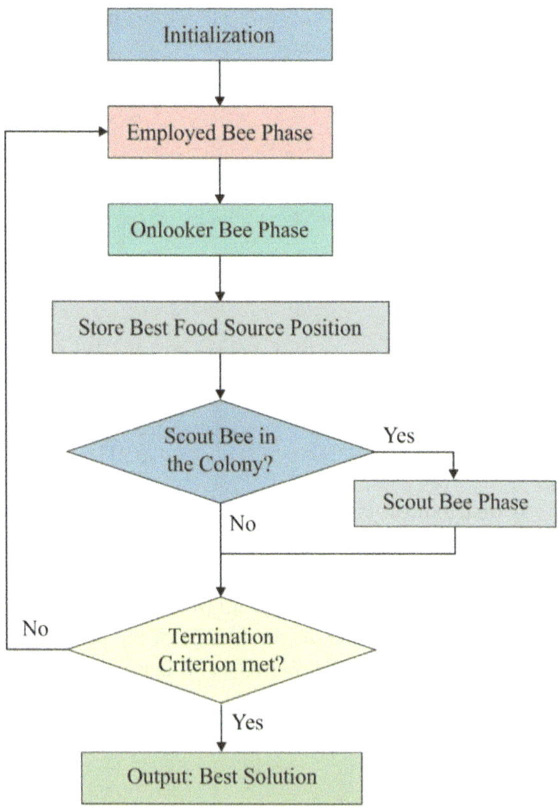

Fig. 2 Flowchart of ABC algorithm

1. Initialization: Create $2N$ locations arbitrarily for determining the location, and progress the employee bees onto the N food sources with maximum nectar numbers.
2. The employed bees search novel food sources from the surrounding using Eq. (4)

$$V_{ij} = x_{ij} + R_{ij}\left(x_{ij} - x_{kj}\right) \tag{4}$$

   where V_{ij} is a novel location, R_{ij} is a arbitrary amount in the range $[-1, 1]$, $k \in \{1, 2, 3 \ldots N\}$ and $k \neq i$.
3. Mark the N food sources with more nectar numbers from the candidate food sources among (1) and (2).
4. Probability P_i is computed as follows:

$$P_i = \frac{f_i t_t}{\sum_{i=1}^{i=N} fit_i} \tag{5}$$

   Here, fit_i is computed utilizing the subsequent equation:

$$fit_i = \begin{cases} \frac{1}{1+f_i} iff_i \leq 0 \\ 1 + abs(f_i) else \end{cases} \tag{6}$$

   where, f_i means the fitness value of the solution.
5. When the solution cannot be enhanced to "limit" trials, it is abandoned. The scout arbitrarily creates a result to changing the old one.
6. Choose the N optimal solutions among candidate results created in step (3) and step (4).
7. Record the optimal result achieved until now and repeat step (2)–(6) till the max cycles.

2.2.2 Multilevel Thresholding

The image threshold models are classified into 2 classes namely, Bi-level thresholding and multilevel thresholding approaches. Initially, Bi-level thresholding is composed of single threshold value that is divided into 2 categories: foreground and background, thus, the image is difficult and comprised with various objects, hence bi-level thresholding is not productive. Secondly, multilevel thresholding techniques were applied for image segmentation. In this approach, a well-known multilevel thresholding is utilized for determining the threshold values, such as, Kapur's entropy. Moreover, the RGB image is composed of 3 typical color components of RGB, and thresholding models are implemented for 3 iterations for computing best threshold values of color component.

2.2.3 Kapur's Entropy

Kapur's model is an unsupervised automated thresholding approach that selects optimal thresholds on the basis of segmented classes. Let $[thl, th_2, th_n]$ be the thresholds that is divided into numerous classes. Next, object function of Kapur's method is expressed by Lang and Jia [15]:

$$H(th_1, th_2, \ldots, th_n) = H_0 + H_1 + H_n \tag{7}$$

$$H_0 = -\sum_{j=0}^{th_1-1} \frac{p_j}{\omega_0} \ln \frac{p_j}{\omega_0}; \omega_0 = \sum_{j=0}^{th_1-1} p_j \tag{8}$$

$$H_1 = -\sum_{j=th_1}^{th_2-1} \frac{p_j}{\omega_1} \ln \frac{p_j}{\omega_1}; \omega_1 = \sum_{j=th_1}^{th_2-1} p_j \tag{9}$$

$$H_n = -\sum_{j=th_n}^{L-1} \frac{p_j}{\omega_1} \ln \frac{p_j}{\omega_1}; \omega_n = \sum_{j=th_n}^{L-1} p_j \tag{10}$$

$H_0, H_1, \ldots, H_n$ is the entropies of various classes, $\omega_0, \omega_1, \ldots, \omega_n$ defines the possibility of a class. To accomplish better threshold values, the FF in Eq. (11) is improved:

$$f_{Kapur}(th_1, th_2, \ldots, th_n) = \text{argmax}\,\{H(thl, th_2, \ldots, th_n)\} \tag{11}$$

It is pointed that the processing complexity of thresholding method results in progressive development when the count of thresholds is enhanced. With certain situations, Kapur's entropy is not efficient for multilevel thresholding. Hence, the ABC relied model with the help of Kapur's entropy is presented to maximize the accuracy and processing speed of thresholding models. The main aim of newly developed technique is to calculate the better threshold values by improving the objective function provided in Eq. (7).

2.3 *Feature Extraction*

The main goal of feature extraction is to acquire image in tiny and exclusive format of matrix vector. It is mainly applied for detecting the dimensionality limitation where image processing depends upon the image classification. It is composed of vanishing input data into a minimized collection of features. Additionally, it is applied for data minimization by estimating the positive features.

2.3.1 Histogram Features

This metric represents the image with respect to pixels. Also, the count of pixels in an image is computed by using a definite value. Converting the power values of histogram which has to be mapped with predefined histogram. From the input image, overall gray levels are computed by the histogram techniques [16]. In this model, 256 Gy levels are ranged from 0 to 255. It is composed of mean, variance, standard deviation (SD), skewness and kurtosis,

Variance: This metric provides the count of gray level fluctuations. The statistical distributions such as difference in the length of lines to the limited extent, which has been applied for differentiating minimum profile in the texture.

Mean: It offers the maximum gray level of images and applicable for harsh idea of power.

SD: The description of SD defines square root of the variance which indicates the image contrast. It is estimated by maximum and minimum variance values. It refers an higher contrast image would have maximum variance while lower contrast image have least variance.

Skewness: It is determined on the basis of histogram. The tail of histogram value is classified as, positive and negative Kurtosis: It is referred as a value of the feasible distribution of arbitrary parameter and represents the image uncertainty. Kurtosis and Skewness were applied in statistical analysis for acquiring the distribution structure.

2.3.2 Texture Features

These features were obtained from input image which is placed subsequent to histogram features. As the abnormality is distributed, the textural representations of all classes are supplemented to reach optimal classification accuracy. The GLCM shows a statistical model which considers the spatial association of pixels.

The GLCM function refers the image texture, by calculating the occurrences of pixel with similar values. Typically, such features were determined by applying GLCM possibility measures and it is acquired from 22 features CT images.

$$G_{P_{ij}=F_{ij/\Sigma_{i,j=0}^{L-1}F_{ij}}} \tag{12}$$

From the above function, F_{ij} implies the frequency of occurrences among 2 Gy levels', L denotes count of determined gray levels, 'i' and 'j' for the provided displacement vector for particular window size.

Energy: The higher constant values are present in gray level distribution which has been evolved from higher dimension.

Entropy: It signifies the data quantity required for compression task. The images which have minimum entropy show lower contrast as well as maximum image pixels in the allocated values.

Homogeneity: It is named as contrast minute that estimates the homogeneous image by considering the common rates for minor gray tone that modifies in paired units.

Contrast: It determines the spatial recurrence of an image as well as differing moments of GLCM. It shows the variance among higher base values of a adjacent pixel organization.

Correlations: It determines the linear basis of gray levels of combined pixels. The observation of digital image correlation refers the optical phase that does not apply the tracking for measuring the image difference.

2.4 RBM Based Classification

RBM is defined as 2-layer rehashed neural approach where the stochastic twofold sources of data were linked with stochastic paired that provides equally-weighed similarities. The manufacturing phase is depicted for removed and expanded RBM in condition (13). It is coursed by RBM that shows confabulation of resourceful data.

$$F(w,h) = -\sum_{i=1}^{i}\sum_{j=1}^{j} I_{ij} w_i h_j - \sum_{i=1}^{I} \alpha_i w_i - \sum_{j=1}^{J} \beta_j h_j \tag{13}$$

where I_{ij} implies the symmetric communication among the clear unit w_i as well as hidden unit h_j, α, β are the bias terms, i, j defines the number of visible and hidden units.

3 Performance Validation

In this section, a detailed set of experimental validation takes place for ensuring the effective performance of the OS-RBM model. The applied ISIC skin cancer image dataset includes 318 images under seven different classes [17]. A total of 21 images in Angioma, 46 images in Nevus, 41 images in Lentigo NOS, 68 images in Solar Lentigo, 51 images in Melanoma, 54 images in Seborrheic Keratosis and 37 images in Basal Cell Carcinoma classes. Figure 3 depicts the sample set of images from the applied dataset.

Figure 4 visualizes the results offered by the proposed OS-RBM model. The figure clearly defined the preprocessed image, masked image, segmented and classified images. It is also evident that the input dermoscopic image is effectively classified into the class type of BCC.

Table 1 showed the comparative analysis of the results offered by the OS-RBM with existing models under diverse measures.

Figure 5 depicts the sensitivity analysis of the OS-RBM with compared models

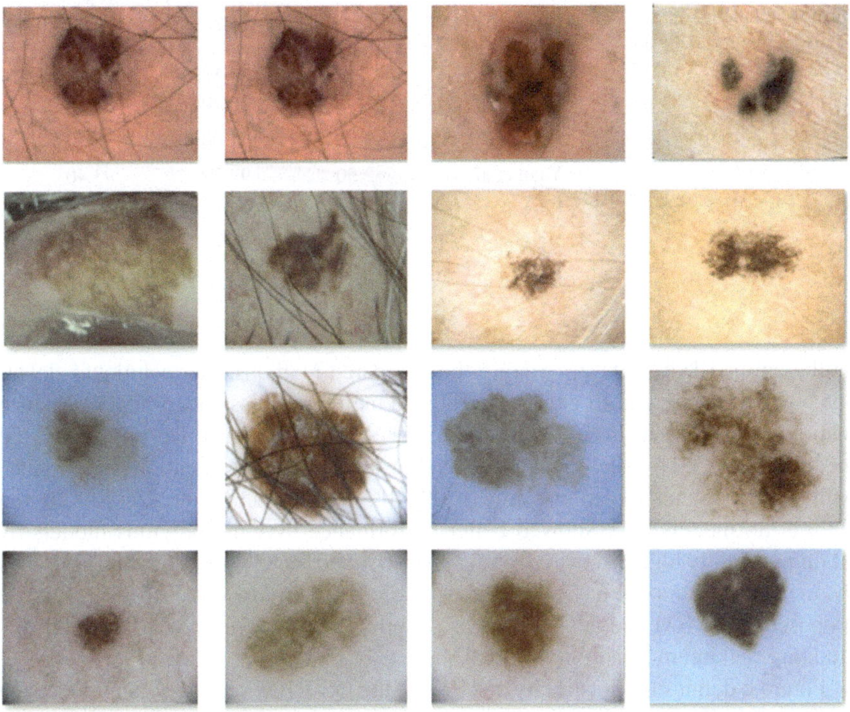

Fig. 3 Sample test images

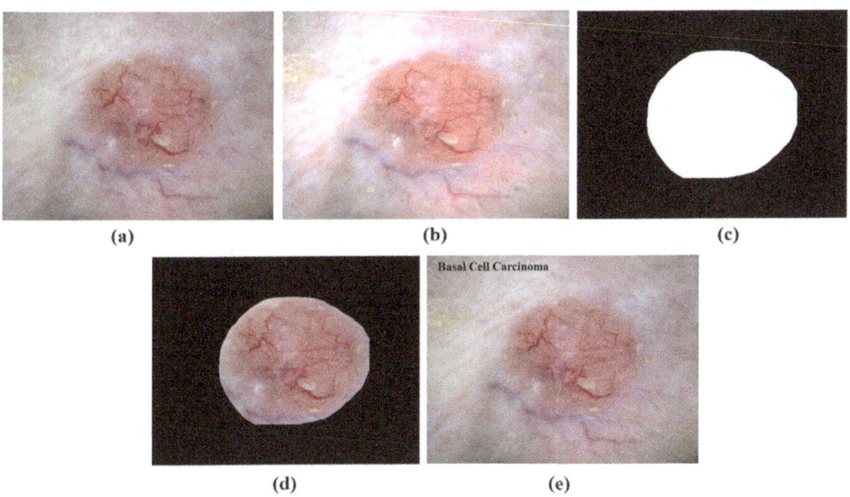

Fig. 4 **a** Original image, **b** Preprocessed image. **c** Masked image. **d** Segmented image. **e** Classified image

Table 1 Comparison of proposed method with state of art methods

Classes	Sensitivity	Specificity	Accuracy
OS-RBM	96.43	97.95	95.68
VGG-19	95.00	68.00	81.20
ResNet-50	90.00	61.00	75.50
Yuan et al.	82.50	97.50	93.40
Li et al.	82.00	97.80	93.20
AI-Magni et al.	85.40	96.69	94.03
Halil et al.	90.82	92.68	93.39

on the applied ISIC dataset. Among the compared methods, the techniques developed by Yuan et al. and Li et al. have showed ineffective classification performance with the minimal sensitivity values of 82.50% and 82% respectively. Besides, the method developed by AI-Magni et al. has showed optimal classification performance with slightly higher sensitivity value of 85.40%. At the same time, ResNet-50 and Halil et al. models have showed better performance over the previous models by attaining slightly higher sensitivity values of 90% and 90.82% respectively. Followed by, the VGG-19 model has offered near optimal sensitivity value of 95%. But the proposed OS-RBM model has shown superior results to compared methods by attaining maximum sensitivity of 96.43%.

Figure 6 defines the specificity analysis of the OS-RBM with alternate models on the provided ISIC dataset. When compared with other approaches, the ResNet-50 and VGG-19 methodologies have exhibited worst classification performance with the least specificity values of 61 and 68% correspondingly. Simultaneously, the model deployed by Halil et al. and AI-Magni et al. showcased optimal performance across earlier models by reaching slightly better specificity values of 92.68 and 96.69% correspondingly. Then, the methods established by Yuan et al. and Li et al. have

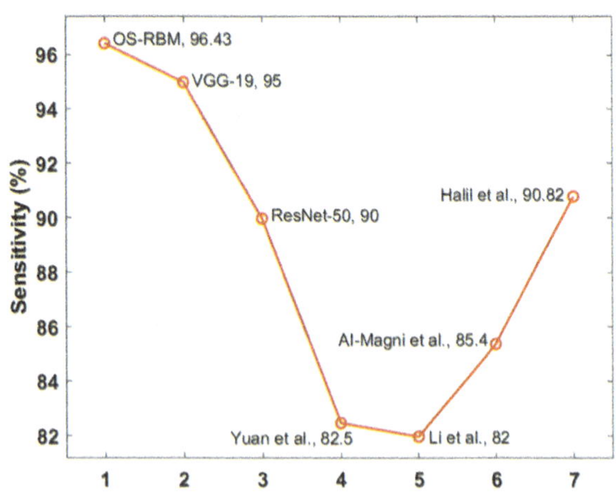

Fig. 5 Sensitivity analysis of OS-RBM method with other models

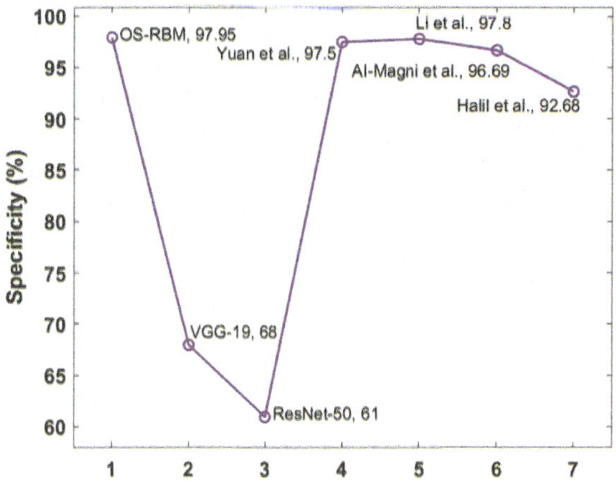

Fig. 6 Specificity analysis of OS-RBM method with other models

provided closer optimal specificity values of 97.50% and 97.80% respectively. Hence, the presented OS-RBM model has implied supreme results than compared methods by achieving maximum specificity of 97.95%.

Figure 7 demonstrates the accuracy analysis of the OS-RBM with previous models on the provided ISIC dataset. Among all other methods, the ResNet-50 and VGG-19 models have depicted ineffective classification performance with the least accuracy values of 75.50 and 81.20% correspondingly. Concurrently, the technique implied by Halil et al. and Li et al. have showcased better performance over the existing models by reaching moderate accuracy values of 93.20% and 93.39% respectively. Next, the methods established by Yuan et al. and AI-Magni et al. have produced closer optimal accuracy values of 93.40% and 94.03% respectively. Thus, the presented OS-RBM model has exhibited qualified results over alternate methods by accomplishing better accuracy of 95.68%.

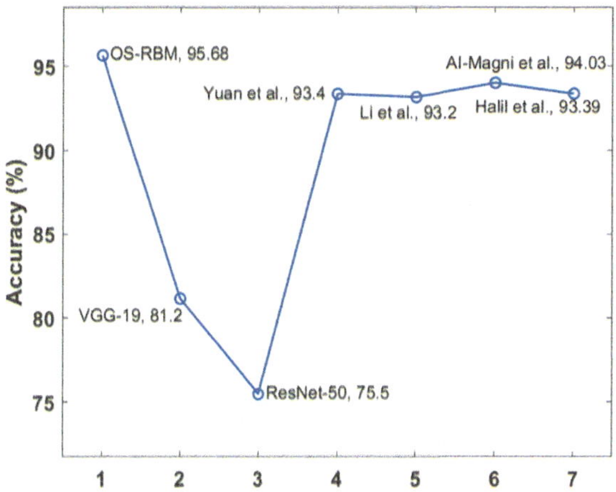

Fig. 7 Accuracy analysis of OS-RBM method with other models

4 Conclusion

This study has introduced an IoMT based skin lesion detection and classification using OS-RBM model. The input dermoscopic images are initially captured by the IoMT devices. Then, GF is applied to eliminate the unwanted details and enhance the image quality. Then, ABC-KR based OS technique is executed to segment the pre-processed images. Simultaneously, the histogram and texture features exist in the segmented image are extracted and finally classified by RBM model. A detailed simulation analysis takes place for ensuring the better outcome of the OS-RBM model using the benchmark dermoscopic images. The experimental outcome ensured the effective classification performance of the OS-RBM model with the maximum sensitivity of 96.43%, specificity of 97.95% and accuracy of 95.68%.

References

1. Basatneh, R., Najafi, B., Armstrong, D.G.: Health sensors, smart home devices, and the internet of medical things: an opportunity for dramatic improvement in care for the lower extremity complications of diabetes. J. Diab. Sci. Technol. **12**(3), 577–586 (2018)
2. Diepgen, T.L., Mahler, V.: The epidemiology of skin cancer. Br. J. Dermatol. **146**, 1–6 (2002)
3. Rigel, D.S., Friedman, R.J., Kopf, A.W.: The incidence of malignant melanoma in the United States: issues as we approach the 21st century. J. Am. Acad. Dermatol. **34**(5), 839–847 (1996)
4. Baldwin, L., Dunn, J.: Global controversies and advances in skin cancer. Asian Pac. J. Cancer Prev. **14**(4), 2155–2157 (2013)
5. Freedberg, K.A., Geller, A.C., Miller, D.R., Lew, R.A., Koh, H.K.: Screening for malignant melanoma: a cost-effectiveness analysis. J. Am. Acad. Dermatol. **41**(5), 738–745 (1999)
6. Lin, J., Han, S., Cui, L., Song, Z., Gao, M., Yang, G., Fu, Y., Liu, X.: Evaluation of dermoscopic algorithm for seborrhoeic keratosis: a prospective study in 412 patients. J. Eur. Acad. Dermatol. Venereol. **28**(7), 957–962 (2014)
7. Kasmi, R., Mokrani, K.: Classification of malignant melanoma and benign skin lesions: implementation of automatic ABCD rule. IET Image Proc. **10**(6), 448–455 (2016)
8. Polevaya, T., Ravodin, R., Filchenkov, A.: Skin lesion primary morphology classification with end-to-end deep learning network. In: International Conference on Artificial Intelligence in Information and Communication (ICAIIC) 2019, pp. 247–250 (2019). 10.1109/ICAIIC.2019.8668980
9. Rahman, M.M., Bhattacharya, P.: An integrated and interactive decision support system for automated melanoma recognition of dermoscopic images. Comput. Med. Imaging Graph. **34**(6), 479–486 (2010)
10. Fong, Y., Kemeny, N., Paty, P., Blumgart, L.H., Cohen, A.M.: Treatment of colorectal cancer: hepatic metastasis. In: Seminars in Surgical Oncology, vol. 12(4), pp. 219–252. Wiley, New York (1996)
11. Abuzaghleh, O., Barkana, B.D., Faezipour, M.: Noninvasive real-time automated skin lesion analysis system for melanoma early detection and prevention. IEEE J. Transl. Eng. Health Med. **3**, 1–12 (2015). https://doi.org/10.1109/JTEHM.2015.2419612
12. Do, T., Hoang, T., Pomponiu, V., Zhou, Y., Chen, Z., Cheung, N., Koh, D., Tan, A., Tan, S.: Accessible melanoma detection using smartphones and mobile image analysis. IEEE Trans. Multimedia **20**(10), 2849–2864 (2018). https://doi.org/10.1109/TMM.2018.2814346
13. Sahu, P., Yu, D., Qin, H.: Apply lightweight deep learning on internet of things for low-cost and easy-to-access skin cancer detection. In: Medical Imaging 2018: Imaging Informatics

for Healthcare, Research, and Applications, vol. 10579. International Society for Optics and Photonics, pp. 1057912 (2019). 10.1117/12.2293350
14. Hosny, K.M., Kassem, M.A., Foaud, M.M.: Skin cancer classification using deep learning and transfer learning. In: 2018 9th Cairo International Biomedical Engineering Conference (CIBEC), 2018, pp. 90–93 (2018). 10.1109/CIBEC.2018.8641762
15. Lang, C., Jia, H.: Kapur's entropy for color image segmentation based on a hybrid whale optimization algorithm. Entropy **21**(3), 318 (2019)
16. Lakshmanaprabu, S.K., Mohanty, S.N., Shankar, K., Arunkumar, N., Ramirez, G.: Optimal deep learning model for classification of lung cancer on CT images. Fut. Gener. Comput. Syst. **92**, 374–382 (2019)
17. https://isic-archive.com/

An IOT Based Medical Tracking System (IMTS) and Prediction with Probability of Infection

Amit Sinha, Ashwin Perti, and Suneet Kumar Gupta

Abstract The size and density of the world population have been increased over the last two decades. Such demographic trends have significant implications for almost all areas of the society, particularly in health and medical sectors. The Life expectancy is increased specially in developing nations where people want a better live with no disease. This chapter aims at developing an IoT based medical appliance called IMTS which works on various integrated hand held medical devices such as pulse sensor, thermal sensor and Oximeter etc. followed by a data analytic experiment over the data received from all such devices corresponding to any patient. The device uses sensors that are interfaced with controller Arduino board. Wireless data transmission done by Arduino through Wi-Fi module. The LED and alarm system give us an easy access to this device. This chapter discusses the disease when we come across to infected people. The idea is to check at personal level through various handheld medical devices and get a probability of infection into you. The pandemic like COVID-19 where problem of running nose, sneezing, cough and fever is the main symptoms, its infection must be controlled. If we are having such appliances that can check our health on daily basis and correlate with my previous health issues and other attributes, the disease can be controlled at initial stage and can stop its infection among family members. The data and result show that the proposed system is a handy tool at domestic level and carrying a good concept.

Keywords IMTS · Arduino board · COVID19 · Pulse sensor · Thermal-sensor

---

A. Sinha (✉) · A. Perti
ABES Engineering College, Ghaziabad, India
e-mail: amit.sinha@abes.ac.in

A. Perti
e-mail: ashwinperti@abes.ac.in

S. K. Gupta
Bennett University, Greater Noida, India
e-mail: suneet.gupta@bennett.edu.in

A. E. Hassanien et al. (eds.), *Cognitive Internet of Medical Things for Smart Healthcare*, Studies in Systems, Decision and Control 311,
https://doi.org/10.1007/978-3-030-55833-8_13

1 Introduction

1.1 Introduction to the Work

For the last two decades, we have been observing several diseases surrounding us. The reason may be our lifestyle, working style, stress level or the contact with our fellow members. The Hospitals and other medical facilities are available to handle such diseases. However, we generally ignore initial stages of any problem. Many persons ignore the initial stage and avoid going to the hospital. They do not take proper consultation to cure it.

This chapter discusses the disease when we come across to infected people. The proposed idea will start from a family and it grows to the society. The idea is to check at personal level through various handheld medical devices and get a probability of infection into oneself and among family members [1, 2]. The chapter considers the current situation of COVID-19 and observing the problem of running nose, sneezing, cough and cold. The reason of taking such symptoms is to its tendency to spread when we come across to any infected person. The government and medical services provide various advisory and precaution methods, still we need an automatic system that monitors and tracks our daily symptoms and its effect on our body and other family members [3, 4].

There are various security and reliability issues that will be required during the development of this medical tracking system. These issues may be confidentiality, Integrity, availability and accountability etc. We tried to include all such issues in the proposed model to give a better interface to the users.

The main objective of this chapter is to observe the Internet of Things (IoT) in Medical science with the use of various handheld medical devices. This gives a new dimension to medical science. This will give Data Analytic report, prediction of diseases with percent of infection and cure human life.

The chapter is divided in four sections viz; Introduction, Requirement of IoT based Medical Tracking Systems, Proposed Medical Tracking System and Inference. Each section is divided into sub sections. Section 1 explains about the concepts, need and applications of IoT, some literature survey and some medical appliances used at domestic level. Section 2 contains the use of IoT devices in medical science. It also describes the architecture of IoT based appliances. Section 3 tells about the proposed work and the integration of all existing appliances. It identifies features and attributes on which some rules are created. The implementation of rules will predict the probability of infection and cure human life [5]. Section 4 discusses on results and the future work in this direction. We consulted the relevant journal papers and included their work in the chapter. All such papers are cited and mentioned in Reference section.

1.2 Literature Survey

In the current scenario of Covid19 we have developed various modern health care system introduces which are based on the technologies like wearable devices using cloud of things and IoT as discussed by author in [6–8]. By using the Internet of Things (IoT) remotely the data can be recorded of the patients and that data can be monitored remotely. For this connection, there is a need of secure data transmission as discussed by author in [1]. Also we have tried to maintain the privacy of the data which is the main concern for our work. The proposed system introduces security of health care and cloud of things. Our major system works in the two parts first is the storage and secondly, the data retrieving stage. In storage stage, data is stored remotely, and updated for future use. In data retrieving stage, retrieve data from cloud.

The authentication over the cloud server can be as per the user request. A patient wearing the wearable devices continually updates the record every 5–10 min. We can also update every minute in case of emergency mode. The wearied device will send results to Mobile-phone either using Bluetooth connection or some other technology like NFC. This can able to give to cloud server using GSM and 4G as used by the author in the paper [9].

At cloud server, each patient is defined with unique address. So data at cloud can authenticate the right patient and provide the required request as done by the author in [2].

1.3 Medical Appliances and Its Uses at Domestic Level

There are various hand-held medical appliances available in the market. These appliances can be used by any person because of their easy to use facility and user friendly nature and also discussed by the author in [10, 11]. Some of such appliances considered for the proposed chapter are:

1. **Pulse Sensor**: This device is used to measure the pulse rate. The designed is such that as the finger is placed over the sensor, it will feed the heart-beat in the form of analog output. It starts working; LED on will starts blinking with each heartbeat. As the output pin of the sensor is connected with the controller, the output can be seen over it. The working principle of sensor is based on light modulation by blood flow through nerves at each heart pulse (Fig. 1).
2. **Thermal Sensor**: This appliance is used for measuring the body temperature of a person. The normal body temperature is 98.6 °F (37 °C), however it can be put in a range depending on the environment and weather of the surroundings. The appliance uses a sensor that is put in contact with body and it senses body temperature. Since it has low self-heating capability, it doesn't require external calibration. A thermal camera locks onto a person's face. Then, the system software locates the hottest point on the face—the corner of either eye, where the upper and

Fig. 1 Pulse sensor

lower eyelids meet—and captures the person's precise body temperature, which appears on an accompanying display (Fig. 2).

3. **Blood Pressure Machine**: This device is mostly used in houses and also discussed by the author in [12]. The digital BP machine monitors the BP. It works on the oscillo-metric principle to measure blood pressure of a human (Fig. 3).

Fig. 2 Thermal sensor

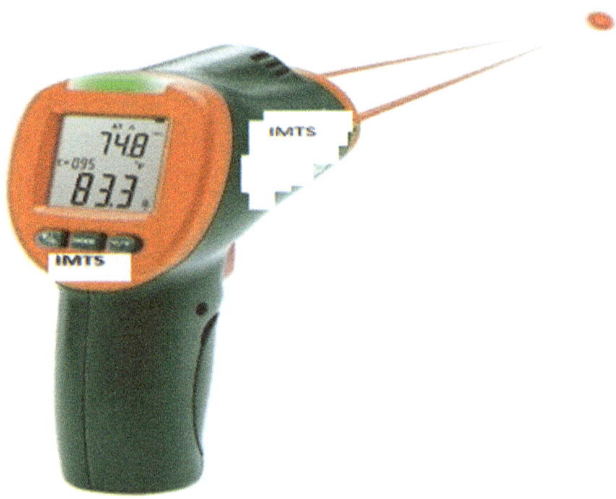

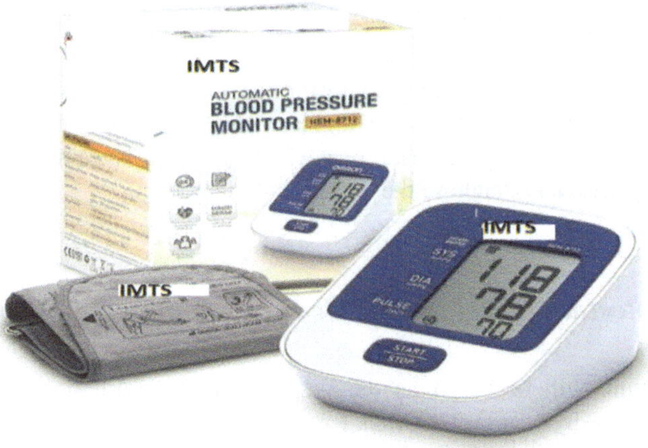

Fig. 3 BP machine

Fig. 4 Oximeter

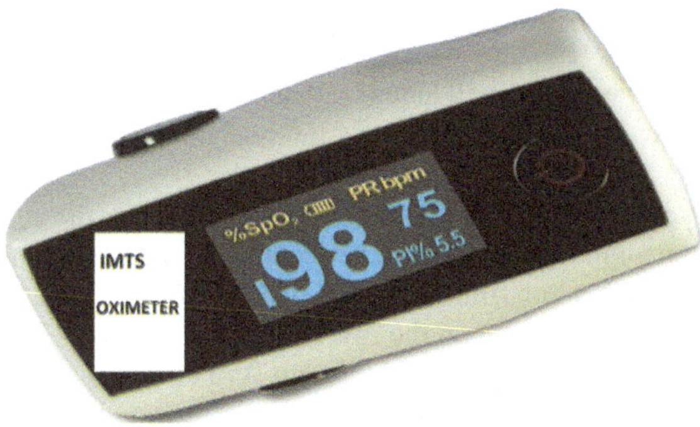

4. **Oximeter**: A sensor in Oximeter is used to measure oxygen saturation in the blood. So haemoglobin saturation is measured by the saturation of peripheral oxygen and that can be measured non-invasively, as example, fixing a clip on the fingertip or ear-tip. The sensor generally employed with a pair of LEDs facing a photodiode that can be used to measure the amount of light passing through the skin (Fig. 4).

2 Requirement of IoT Based Medical Tracking Systems

The present chapter works on various handheld medical devices generally used in houses also discussed in [13], following are some of the requirements for the proposed system IMTS:

2.1 IoT and Medical Science

IoT and Medical Science have been playing the key role in identifying and controlling the various health factors of a human-being. It exists from measuring the temperature of a human being to measure the heart beat which are one of the essential factors for a doctor to analysis and diagnose the disease in human being.

2.2 Architecture of IoT Based Medical Appliances

The architecture will be based on the Wearable IoT-cloud-based Health Appliances Monitoring System (WIHAMS), for real-time personal health monitoring as discussed by the author in [14]. The WIHAMS adopts the BAHASN (body area health appliance sensor network) framework in the support of real-time health appliance monitoring [15]. Several wearable health appliance sensors have been embedded [16, 17], including the Oximeter, pulse sensing, heartbeat, body temperature, and the blood pressure sensors. Secondly, the majority of existing wearable health monitoring systems requisite a smart phone as data processing, visualization and transmission gateway, which will indeed impact the normal daily use of the smart phone. For quick view of the real-time data a lightweight wearable LCD can be easily embedded to get an alternative solution.

The model we are proposing also works for the current scenario of Covid19. The same model is applicable to identify the person is infected or not also discussed by the author for similar type of architecture in [18–21] and can take the appropriate action depending on the severity of the disease.

Data will be collected through Wi-Fi devices and send to the Cloud device for the data storage and the data-mining part. Through this, we will be getting the required data, which will be analyses and send to the respective user through the Mobile App like ArogyaSetu App. The architecture shown in Fig. 5 gives an idea of working of proposed model.

In Fig. 5, with the help of Arduino the data will be sent to various devices like LCD, RFID and Wi-Fi enabled devices. The data will be received from various domestic health sensor devices and coordinated with the other devices like RFID and LCD with Wi-Fi device to connect with the cloud for storage purpose. With the advancement of cloud computing technology, several functions can be implemented in the IMTS.

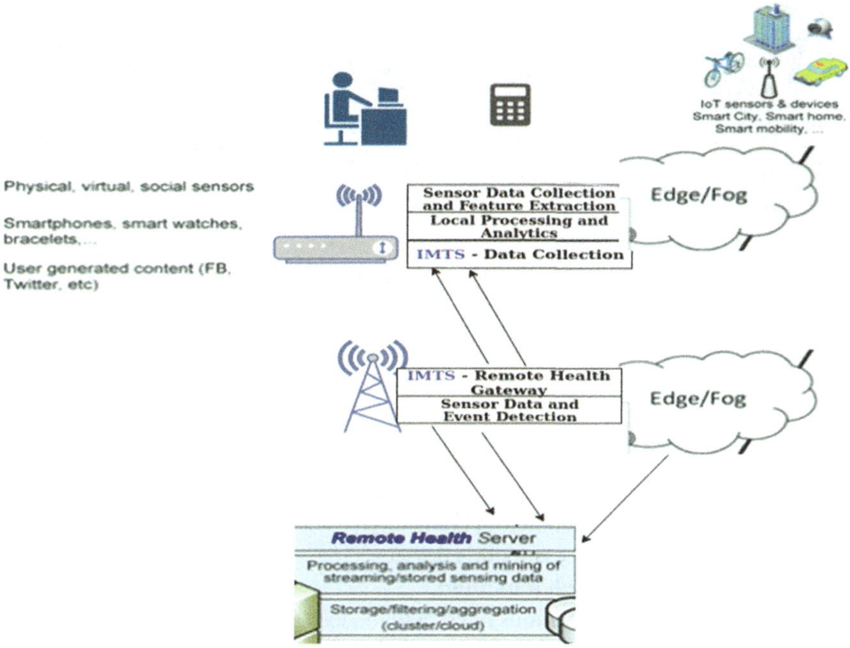

Fig. 5 Architecture of wearable IoT based health appliance monitoring system (WIHAMS)

3 Proposed Medical Tracking System

3.1 Basic Model Using Integration of Available Appliances

The work proposes a tracking system for the persons where we observe some medical changes or health issues. This will help a family to keep an eye on its members and take a precautionary measure well in advance. The second phase of the model is to observe other family members on the basis of similar features/attributes. It can protect whole family members. The chart shown in Fig. 6, can demonstrate the idea.

In short, the model can be described among three main points:

1. **Integration of various hand held medical appliances**
   The connected appliances measure the relevant values of a person and sent to the interface device. In other words, thermal check meter measures the temperature of a family member and sends to central device, while BP machine checks the blood pressure, oximeter will measure the oxygen flow in our body and pulse sensor checks the pulse rate and transfers the value to the central point.
2. **Collection of all data through an interface (IMTS)**
   The concept of central interface device is IoT based Medical Tracking system (IMTS). The idea to develop IMTS is to collect values sent by all connected appliances. The proposed device will use the features of Bluetooth technology and

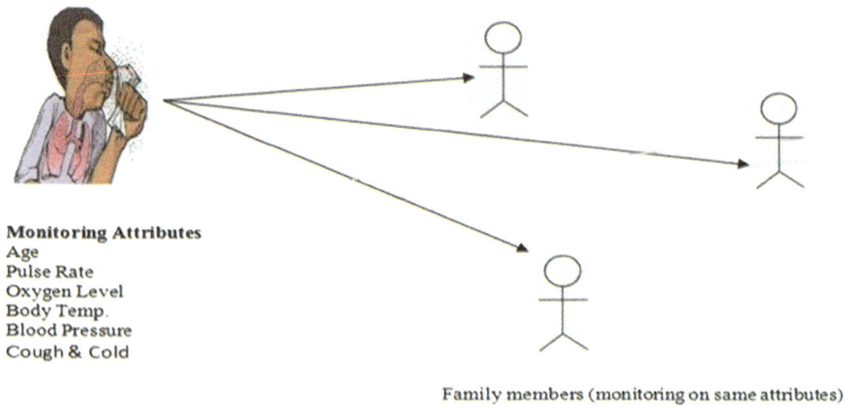

Fig. 6 Study of disease and its infection to family members

wireless technology. The internet connectivity with IMTS will also be available that give a variety of accessing this device.

3. **Analysis of data through a method**
   The data received from various hand held devices are stored and passed through a specified method. This method treats all these data as an attributes. Several other attributes are also required from users or patients so that we can process and give a better result to them.

3.2 *Feature Selection*

We are proposing an appliance for a general purpose can be used easily at domestic level. Also, this appliance (IMTS) and its output give an alarm towards healthy body. It is not a replacement of proper check-up done by Doctors and Para-medical staffs. The tracking system will monitor our day to day life, store all values of our BP, Pulse rate, Oxygen level and body temperature at-least twice a day especially for asymptotic patients. Here, we have identified following features and their normal value. These are shown in Table 1.

We kept this list limited; however it can be extended as per the severity and for future work.

3.3 *Methodology and Experimental Design of Proposed System*

The IMTS is more inclusive in its approaches and methods are based on many existing disciplines. Also, the system is at preliminary stage and is growing; it has

Table 1 Features/attributes considered for IMTS

Appliances/attributes	Normal value	Reason to select this feature/attribute
Pulse sensor/pulse rate	60–100 beats/min	It varies with cough and cold or running nose
Thermal sensor/temperature	98.6 °F	It is directly proportional to any disease. It will increase when the problem increases
BP machine/BP value	120–80 mm Hg	It may have low BP or high BP
Oximeter/O_2 value	90–100 mm Hg	It decreases in case of cough and cold
Feeling of cold	Input by person	An important part of our work
Feeling of cough	Input by person	Infection can be transferred to other family members. Earlier diagnosis is very important
Last 3 times disease	Input by person	A mandatory input, because other symptoms may play an important role in prediction of infection
Age of person/patient	Input by person	An important input because we can overcome from infection by our strong immune system

no restriction and open to include many attributes and scope of improvement. The medical health system is a broader area of research that covers different aspects of health monitoring system such a utilization, person-to-person spread and demand and supply trend. The experimental design is a network of various devices with Wi-Fi features and cloud technology. It is shown in Fig. 7.

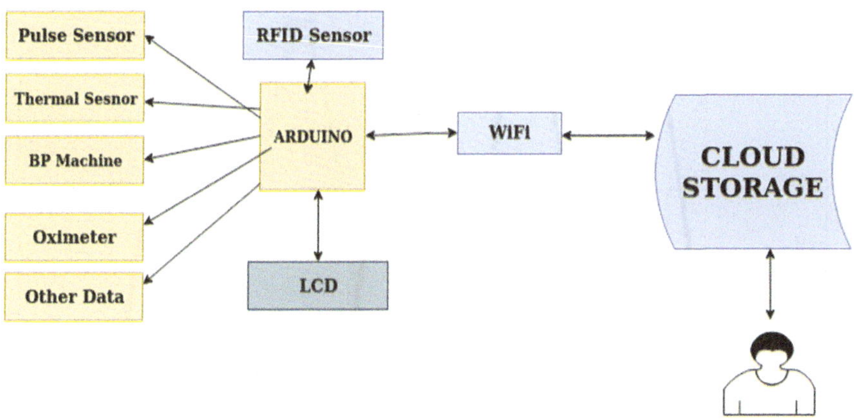

Fig. 7 Integration of hand held medical devices with IoT based appliances

3.4 *Rule Based Classifiers to Predict the Probability of Infection*

We have considered all features and its values. However, the normal values are described in Table 1; it is not that alarming in all cases at domestic level. If one value increases, it should be correlated with other attributes [3]. In other words, we can check if a person is of age between 20 and 30 years and has no previous symptoms, the body temperature of that person increases upto 1 °F, has no immediate worry.

Therefore, the combination of all values of all attributes must be considered and analysed properly. On this basis, some rules have been created. These rules are created with the consultation with few doctors and Para-medical staffs. Also a regular monitoring of some persons, observing all values, storing and then running through the designed rules give a better idea. The known results are validated through the proposed rule and it is found a satisfactory result. The rules with their dependencies on spreading infection are mentioned below:

The process of describing crisp values into linguistic terms is called fuzzy sets also discussed in [4] and the process is called Fuzzy-fication. Firstly, the input (Heart Rate) HR is fuzzified into "Very Low", "Low", "Normal", "High", and "Very High" type-1 membership functions as shown in Fig. 8.

Each point in HR range may have one or more non-zero membership value; such as HR = 63 has non-zero membership value in two membership functions, i.e. 'Normal' and 'Low'.

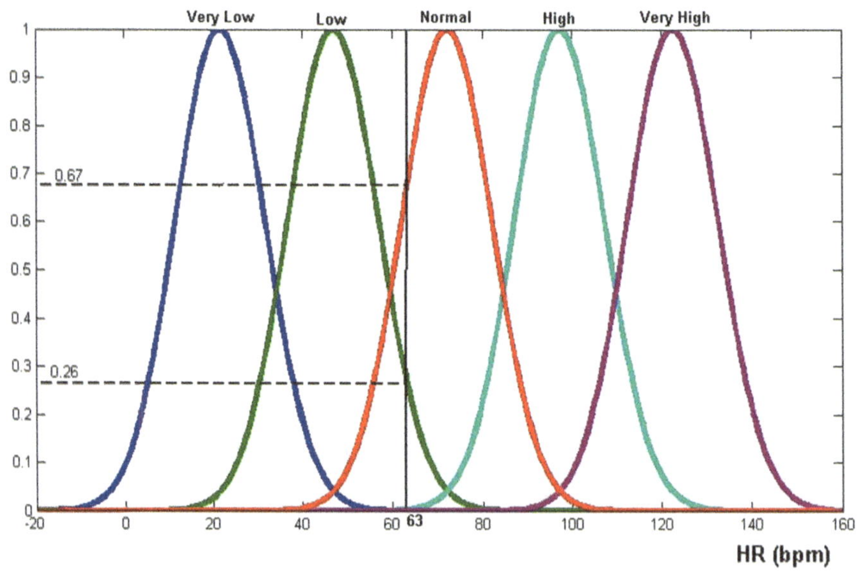

Fig. 8 Type-1 membership functions for heart rate (HR)

The degree of membership is also referred to as the "truth" value. The HR = 63 will be assigned fuzzy set with higher membership grade, i.e. Normal in this example.

Secondly, the output mean blood pressure is represented into "Very Low", "Low", "Normal", "High", and "Very High" type-2 membership functions as shown in Fig. 9.

FLS Rules are extracted from the numerical data depend on the conducted experiment, to map input with output. Each rule has to parts; "IF" part; which is called antecedent and "Then" part; which is called consequent (Table 2).

The extracted FLS Rules are given below;

Rule 1: IF HR is "Very Low" THEN BP is "Very Low"

Rule 2: IF HR is "Low" THEN BP is "Low"

Rule 3: IF HR is "Normal" THEN BP is "Normal"

Rule 4: IF HR is "High" THEN BP is "High"

Rule 5: IF HR is "Very High" THEN BP is "Very High".

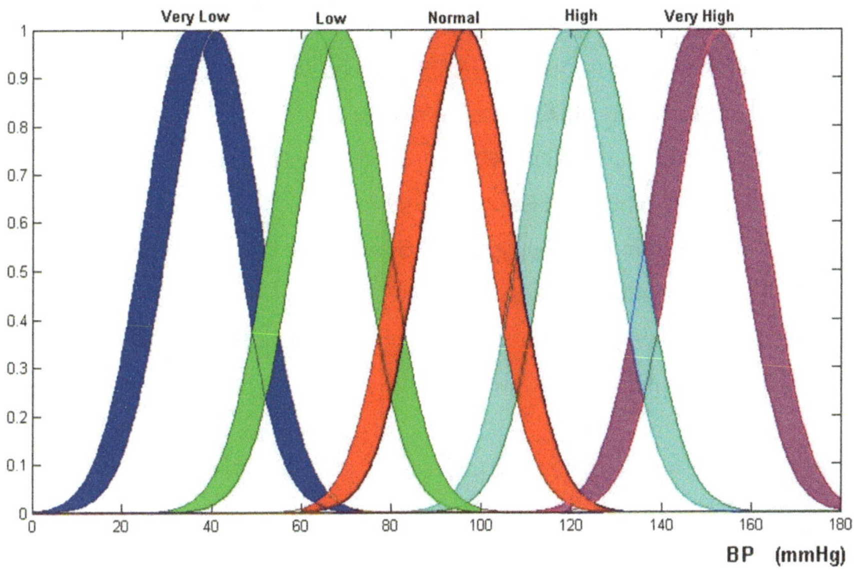

Fig. 9 Type-2 membership functions for heart rate (HR)

Table 2 Proposed range of all features with membership classification

Pulse rate (PR) (per min)	Membership	Temperature (°F)	Membership	Age	Membership	Probability of disease
<58	Excellent	X = 98.6	Normal	<18 years	Child	Low
59–63	Good	98.6 < X < 100	Medium	18–50	Adult	Medium
64–70	Fair	X > 100	High	>50	Senior	High

The extracted FLS Rules are given below;

Rule 1: IF Temp is "Low" & PR is "Excellent" & age is "Child"
Then Probability of Disease is "Very Low"
Rule 2: IF Temp is "Medium" & PR is "Excellcnt" & age is "Adult"
Then Probability of Disease is "Very Low"
Rule 3: IF Temp is "Medium" & PR is "Excellent" & age is "Senior"
Then Probability of Disease is "Very Low"
Rule 4: IF Temp is "Medium" & PR is "Excellent" & age is "Child"
Then Probability of Disease is "Very Low"
Rule 5: IF Temp is "Medium" & PR is "Excellent" & age is "Senior"
Then Probability of Disease is "Medium"
Rule 6: IF Temp is "Medium" & PR is "Good" & age is "Child"
Then Probability of Disease is "Very Low"
Rule 7: IF Temp is "High" & PR is "Good" & age is "Child"
Then Probability of Disease is "Medium"
Rule 8: IF Temp is "High" & PR is "Good" & age is "Senior"
Then Probability of Disease is "High"
Rule 9: IF Temp is "High" & PR is "Fair" & age is "Child"
Then Probability of Disease is "High" (1)

4 Inference

4.1 Results and Conclusion

This book chapter gives an idea about some diseases such as cough and cold can be monitored at domestic level. The symptoms are clearly seen by the persons themselves and by family members. It also affects various other bodies functioning such as, in case of cough and Cold, the oxygen value decreases due to breathing problem. The body temperature may also increase with high pulse rate. We have identified cough and cold due to its probability to infect others.

Therefore, an IoT based medical Tracking system is needed that tells you the current probability of infection with a LCD display.

Table 3 Proposed range of all features with membership classification

Pulse rate	Temp	Oxygen	BP	Age	Cold	Suggestive measure
60–100 beats/min	99	60	60	20	Yes	Contact family doctor
50	98.6 °F	70	70	30	No	Contact family doctor
110	100	120–80 mm Hg	80	40	Yes	After 2 days
120	101	130	90–100 mm Hg	50	No	Quarantine yourself
130	102	140	150	40	Yes	Immediately

We have established an experimental setup with the architecture shown in Fig. 5 and collected the data from different integrated devices. To execute and check the performance of the proposed system we have taken 50 patients/user as a sample data and recorded their information measured by integrated handheld devices twice in a day for a month. A snapshot of ($50 \times 2 \times 30$ days $= 3000$) data in the form of 6 tuple set.

$$< \mathbf{P, T, O, B, A, C} >$$

All these parameters
P—Pulse Rate
T—Temperature
O—Oxygen Level
B—Blood Pressure
A—Age of Patient
C—Cold/Cough (yes/no)

are shown in Table 3.

These datasets are simulated through Python Programming with a function skfuzzy. Now these functions are embedded with the FIS rules in Eq. (1). The snapshot of code is shown in Fig. 10 and membership function in Fig. 11.

The Activation Function for predicting the level of infection is shown in Fig. 12 and its output is shown in Fig. 13.

4.2 *Discussion and Future Work*

The result obtained from the experimental setup is validated through two methods.

1. **Prospective Validation**: In prospective validation we have checked our result with known result and found a significant improvement and equivalent output with the proposed IMTS.

```
# Generate fuzzy membership functions
pulse_lo = fuzz.trimf(x_pulse, [0, 0, 5])
pulse_md = fuzz.trimf(x_pulse, [0, 5, 10])
pulse_hi = fuzz.trimf(x_pulse, [5, 10, 10])

thermal_lo = fuzz.trimf(x_thermal, [0, 0, 5])
thermal_md = fuzz.trimf(x_thermal, [0, 5, 10])
thermal_hi = fuzz.trimf(x_thermal, [5, 10, 10])

bp_lo = fuzz.trimf(x_bp, [0, 0, 13])
bp_md = fuzz.trimf(x_bp, [0, 13, 25])
bp_hi = fuzz.trimf(x_bp, [13, 25, 25])

oxygen_lo = fuzz.trimf(x_oxygen, [0, 0, 13])
oxygen_md = fuzz.trimf(x_oxygen, [0, 13, 25])
oxygen_hi = fuzz.trimf(x_oxygen, [13, 25, 25])

infection_lo = fuzz.trimf(x_Infection, [0, 0, 13])
infection_md = fuzz.trimf(x_Infection, [0, 13, 25])
infection_hi = fuzz.trimf(x_Infection, [13, 25, 25])
```

Fig. 10 Initialization of the parameters for the fuzzy set

2. **Retrospective Validation**: In retrospective validation when this model is used for a new user/patient it also declares a good prediction that is validated that is found well again.

Thus, it will help the society and assist the health sector in both the validations.

The impact of IoT is increasing day by day. The medical field is not escaped and has a significant effect on treatment and help in diagnosis of the disease.

The present work proposed an idea of IoT based medical tracking system. The integration of various portable devices like BP machine, Thermal sensor and Oximeter etc. connected with proposed IMTS, give a better and new dimension of learning to the end user/patient. The analysis of these data gives a good outcome and learning to Doctors and Para-Medical staffs. The final outcome is in the form of prediction and probability of infection. It uses various FLS rules and simulated in MATLAB. We implemented it on nearly 500 users and collected data twice in a day for one month. The data is normalized and simulated on weekly basis. The final outcome is significant and provides a handy tool to all our family members.

This technology becomes more useful when it is combined with the features of mobile computing. The use of mobile computing also extends the functionality of

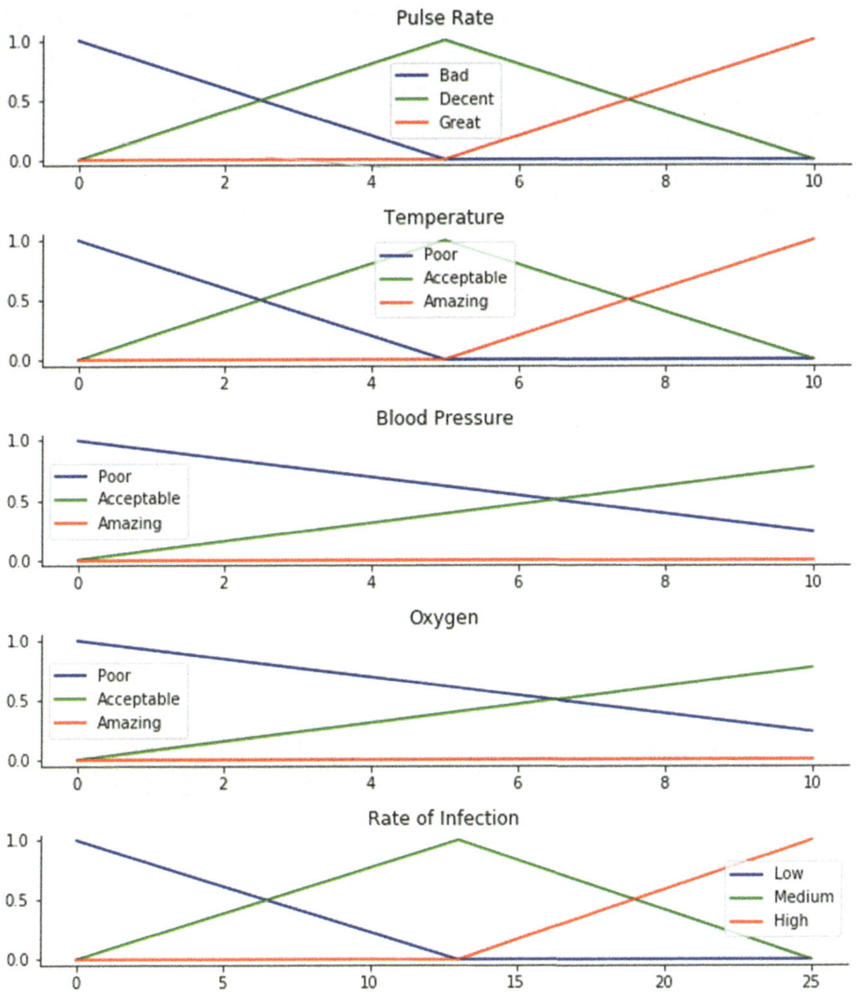

Fig. 11 Membership function

IoT in the case of healthcare devices forming a environment by bringing a massive support in the form of mobile health (m-health). A systematic literature review (SLR) protocol is proposed to study how IMTS assists healthcare systems. The proposed system may be extended by adding more features to it, such as ECG report, Glucose level and breathing problem of the patients. Also, the rules can be analysed again and modified with the use of extended features.

```
infection_activation_lo = np.fmin (active_rule1, infection_lo) # removed entirely to 0

# For rule 2 we connect acceptable service to medium infection
infection_activation_md = np.fmin(pulse_level_md, infection_md)
# For rule 3 we connect high pulse OR high features with high infection
active_rule3 = np.fmax(pulse_level_hi, bp_level_hi)
infection_activation_hi = np.fmin(active_rule3, infection_hi)
infection0 = np.zeros_like(x_Infection)
# Visualize this
fig, ax0 = plt.subplots(figsize=(8, 3))
ax0.fill_between(x_Infection, infection0, infection_activation_lo, facecolor='b', alpha=0.7)
ax0.plot(x_Infection, infection_lo, 'b', linewidth=0.5, linestyle='--', )
ax0.fill_between(x_Infection, infection0, infection_activation_md, facecolor='g', alpha=0.7)
ax0.plot(x_Infection, infection_md, 'g', linewidth=0.5, linestyle='--')
ax0.fill_between(x_Infection, infection0, infection_activation_hi, facecolor='r', alpha=0.7)
ax0.plot(x_Infection, infection_hi, 'r', linewidth=0.5, linestyle='--')
ax0.set_title('Output membership activity')
```

Fig. 12 Code for activation function

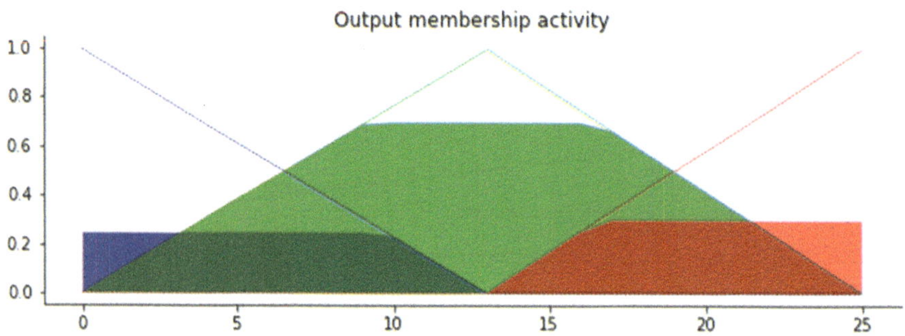

Fig. 13 Output membership activation

References

1. Riazul Islam, S.M., Kwak, D., Kabir, M.H., Hossain, M., Kwak, K.-S.: The internet of things for health care: a comprehensive survey. IEEE Access. https://doi.org/10.1109/TDSC.2015.2406699
2. Masud, M.M., Serhani, M.A., Navaz, A.N.: Resource-Aware Mobile-Based Health Monitoring, pp. 2168–2194. IEEE (2015)
3. Pal, A., Visvanathan, A., Choudhury, A.D., Sinha, A.: Improved heart rate detection using smart phone. In: 29th Annual ACM Symposium on Applied Computing, pp. 8–13. ACM, New York (2014)
4. Mahmood, U., et al.: Type-2 fuzzy classification of blood pressure parameters. In: ISSNIP 2007, pp. 595–600 (2007)
5. Pigadas, V., Doukas, C., Plagianakos, V.P., Maglogiannis, I.: Enabling constant monitoring of chronic patient using android smart phones. In: 4th International Conference on Pervasive Technologies Related to Assistive Environments, pp. 63–1632. ACM, New York (2011)
6. Wan, J., et al.: Wearable IoT enabled real-time health monitoring system. EURASIP J. Wireless Commun. Network. **2018**, 298 (2018)
7. Patil, S., Pardeshi, S.: Health monitoring system using IoT. IRJET **05**(04), 1678–1672 (2018)

8. Dubey, D., et al.: Smart phone applications as a source of information on stroke. J. Stroke **16**(2), 86–90 (2014)
9. Mdhaffar, A., Chaari, T., Larbi, K., Jmaiel, M., Freisleben, B.: IoT-based health monitoring via LoRaWAN. In: IEEE EUROCON 2017 (2017)
10. Al-Hamadi, H., Chen, I.-R.: Trust-based decision making for health IoT systems. IEEE Internet Things J. https://doi.org/10.1109/JIOT.2017.2736446
11. Thangaraj, M., Ponmalar, P.P., Anuradha, S.: Internet of things (IOT) enabled smart autonomous hospital management system—a real world health care use case with the technology drivers. In: 2015 IEEE International Conference on Computational Intelligence
12. Bansal, A., Kumar, S., Bajpai, A., Tiwari, V.N., Nayak, M., Venkatesan, S., Narayanan, R.: Remote health monitoring system for detecting cardiac disorders. IET Syst. Biol. **9**(6), 309–314 (2015)
13. Kumar, M.A., Ravi Sekhar, Y.: Android based health care monitoring system. In: IEEE Sponsored 2nd International Conference on Innovations in Information Embedded and Communication Systems ICIIECS (2015)
14. Guo, W., Xiong, N., Vasilakos, A.V., Chen, G., Cheng, H.: Multi-source temporal data aggregation in wireless sensor networks. Wirel. Pers. Commun. **56**(3), 359–370 (2018)
15. Qiu, Y., Ma, M.: Secure group mobility support for 6LoWPAN networks. IEEE Internet Things J. **5**(2), 1131–1141 (2018)
16. Sinha, R.S., Wei, Y., Hwang, S.H.: A survey on LPWA technology: LoRa and NB-IoT. ICT Express **3**(1), 14–21 (2017)
17. Wan, J., O'Grady, M., Li, M., Alaqi, M.: Time-bounded activity recognition for ambient assisted living. IEEE Trans. Emerg. Top. Comput. **3**(1), 1–14 (2018)
18. Hallfors, N.G., Alhawari, M., Abi Jaoude, M., Kifle, Y., Saleh, H., Liao, K., Ismail, M., Isakovic, A.F.: Graphene oxide: nylon ECG sensors for wearable IoT healthcare—nanomaterial and SOC interface. Analog Integr. Circ. Sig. Process. **1**, 1–8 (2018)
19. Varatharajan, R., Manogaran, G., Priyan, M.K., Sundarasekar, R.: Wearable sensor devices for early detection of alzheimer disease using dynamic time warping algorithm. Clust. Comput. **6**, 1–10 (2017)
20. Romero, L.E., Chatterjee, P., Armentano, R.L.: An IoT approach for integration of computational intelligence and wearable sensors for Parkinson's disease diagnosis and monitoring. Health Technol. **6**(3), 167–172 (2016)
21. Dubey, D., Amritphale, A., Sawhney, A., Amritphale, N., Dubey, P., Pandey, A.: Smart phone applications as a source of information on stroke. J. Stroke **16**(2), 86–90 (2014)

The manufacturer's authorised representative in the EU is Springer Nature Customer Service Centre GmbH, Europaplatz 3, 69115 Heidelberg, Germany. If you have any concerns regarding our products, please contact ProductSafety@springernature.com

Printed and bound by CPI Group (UK) Ltd, Croydon, CR0 4YY
30/09/2026
02225288-0002